TESTOSTERONE

Joe Herbert is Emeritus Professor of Neuroscience and a Fellow of Gonville and Caius College at the University of Cambridge. His areas of expertise include the role of hormones in the ability of the adult brain to make new nerve cells and repair the brain, how hormones regulate behaviour, the neuroscience of stress, how hormones, genes, and the social and psychological environment interact to promote the risk for depression, and studies on the way that hormones and genes influence financial decision-making. He has authored (and co-authored) around 250 scientific papers, and is the author of *The Minder Brain: How your brain keeps you alive, protects you from danger, and ensures that you reproduce* (World Scientific Press, 2007).

JOE HERBERT

testosterone

the molecule behind power,
sex, and the will to win

OXFORD
UNIVERSITY PRESS

OXFORD
UNIVERSITY PRESS

Great Clarendon Street, Oxford, OX2 6DP,
United Kingdom

Oxford University Press is a department of the University of Oxford.
It furthers the University's objective of excellence in research, scholarship,
and education by publishing worldwide. Oxford is a registered trade mark of
Oxford University Press in the UK and in certain other countries

First published 2015
First published in paperback 2017

Impression: 1

Published in the United States of America by Oxford University Press
198 Madison Avenue, New York, NY 10016, United States of America

British Library Cataloguing in Publication Data

Data available

Library of Congress Cataloging in Publication Data

Data available

ISBN 978–0–19–872497–1 (Hbk.)
ISBN 978–0–19–872498–8 (Pbk.)

Printed in Great Britain by
Clays Ltd, St Ives plc

For my friend
T. C. Anand Kumar (1936–2010)
with whom I ate, drank, laughed and argued for more than 40 years.

PREFACE

There are, I suppose, many reasons why professional scientists, who normally write highly technical articles, should be tempted to write a book such as this. Scientists, like painters, musicians, and many others, are obsessed with their subject. This doesn't always make them very good at being members of a family, let alone companions at a dinner table. But one feature of an obsession is the desire to share it with others. Something so fascinating and self-absorbing, the reasoning goes, must surely be as interesting to anybody else. Hence the existence of those pub-bar bores. And yet: science is so central to everybody's life, so omnipresent in our world, so influential on everything we do, that any scientist's urge to tell the world about what he or she does is irresistible. The media encourage such a view: no day passes without a headline story about science of some sort. The growth of professional scientific journalism is testament to the public hunger for science: what is happening and will it affect me?

So if these writers exist, why would someone like me, a scientist and not a journalist, write a book about my subject? A simple reason: someone reporting on a subject is not the same as someone doing it. Journalists are incredible: they pick up a story about which they may know nothing to begin with, quickly and effectively, and write lucidly and informatively about it. But it's not the same. A scientist hasn't thought about his subject for a few days, or weeks, but for years. So a scientist has a point of view: moreover, he/she knows that science is not simple, and there are often many points of view about a particular piece of scientific research. Mulling over your subject results in a state of mind that is not easily reproduced in any other way: a sort of maturation of thought. That doesn't mean that the scientist is necessarily right in his/her views; in fact, one of the important endpoints of this state of mind is the realization of what is not known, and how far what we think we know is incomplete or uncertain. It's also the ability to recognize the next big question. So writing a book such as this is not simply an account of the facts, but an interpretation and an acknowledgment that, in any part of science, there are huge pieces missing from the puzzle. You tell a story, but one full of twists and turns. No simple message or bottom line.

Hormones are fascinating. These chemicals, produced in tiny amounts, exert powerful influences on our lives, and their discovery was a huge landmark in biology and medicine. The fact that they also have powerful effects on the brain make them all the more fascinating, for the brain itself, that crucible of humanity, cannot fail to interest us, who are largely what are brains are. Since our understanding of the brain is so incomplete (a very mild way of putting our ignorance) the interaction between hormones and brains becomes that more intriguing. So I want you to share in my fascination, and I hope I have the skill to enable you to do so. But do not expect a complete story: the gaps are too wide to be disguised. Scientists are sometimes not too good at admitting ignorance (a favourite phrase in the scientific literature is that something is 'not fully understood' which actually means 'we havn't a clue'). Of all the powerful hormones, none is more influential than testosterone, or so I will try and persuade you. We know enough to know that.

I have to take responsibility for my book, but it's been greatly enhanced by my friends who have taken time to read chapters, give me ideas and suggest numerous improvements. They include Alan Dixson, Barry Everitt, Mick Hastings, Barry Keverne, and Scarlett Pinnock, all distinguished scientists and collaborators; Richard Green, Jay Schulkin, and Tirril Harris, luminaries in their field; Jeremy

Prynne, a notable poet and critic; and John Bancroft, who has written the definitive book on sexual disorders. My son Daniel, making his way as a writer in New York, has given me valuable guidance on style and clarity. My other son, Oliver, busy as a young doctor, helps me to stay in touch with clinical matters. Latha Menon and Emma Ma, of OUP, have made the process of editing this book a pleasure and an education. Finally, I have, for most of my career, been immersed in the stimulating environment of the University of Cambridge and my college, Gonville and Caius, where, almost every day, one learns something new; to all these colleagues, friends and acquaintances, I express my thanks and admiration. My ever-patient wife, Rachel Meller, has, as always, tolerated my mental and physical absences with kindness, understanding and support, and the clarity of her writing has been a model.

CONTENTS

LIST OF ILLUSTRATIONS

1

Testosterone and Human Evolution

It takes millions of years to perfect a dramatically new animal model, and the pioneer forms are usually very odd mixtures indeed. The naked ape is such a mixture. His whole body, his way of life, was geared to a forest existence, and then suddenly (in evolutionary terms) he was jettisoned into a world where he could survive only if he began to live like a brainy, weapon-toting wolf.
                                   Desmond Morris (1967), *The Naked Ape*.
                                                      Jonathan Cape, London

Humans are talking primates, but in fact their behavior is not very different from that of chimpanzees. People engage in verbal fights, provocative or impressive word displays, protesting interruptions, conciliatory remarks, and many other patterns of verbal activity that chimpanzees perform without an accompanying text. When humans resort to actions instead of words the resemblance is even greater. Chimpanzees scream and shout, bang doors, throw objects, call for help, and afterward they may make up by a friendly touch or embrace.

F. de Waal (1998), *Chimpanzee Politics*. Revised edition.
Johns Hopkins University Press, Baltimore

We are born into our modern world with a brain that was developed for a more primeval one. Early humans had little control over their world. Their brain had evolved to cope with the exigencies of surviving in that harsh environment. Getting food and water, keeping warm/cool, finding shelter, beating off rivals, avoiding becoming prey: all required the adaptive qualities that Darwin and the renowned neuroscientist Ramon y Cajal recognized as necessary for the struggle for survival.*

Everything we call 'human' depends on the evolution of the human brain. Look at our closest relatives: chimpanzees and gorillas. Their brains look, at a casual first glance, very similar to a human one. But history tells us that this is deceptive. Each night, a chimpanzee builds a sleeping nest. If we were to roll back time for 10,000 years, we would see chimpanzees doing much the same thing. Of course, they can adapt. But chimpanzees (like any other primate) have no technical or cultural history that bears any resemblance to ours. We change our surroundings and the conditions in which we live: we invent tools, machines and agriculture; ensure adequate supplies of easy-to-get food and clean, accessible water. The habitations of humans 10,000 years ago (or even 1,000 years ago) were very different from those of today.[1] Though there are examples of other species doing vaguely similar actions—made much of by those who want to emphasize the commonalities between humans and other species—no other species comes close to man in technical or conceptual ability. Charles Darwin wrote:

* 'From a teleological point of view, we may think of the nervous system as entrusted with several tasks: collecting a large number of external stimuli; classifying them as to kind; and communicating them with great speed, range, and precision to motor systems, while simultaneously minimizing unproductive, diffuse, or inappropriate responses. Moreover, we can see that it has the added responsibility of maintaining the harmony and integrity of the various related parts of the organism by restraining and directing the entire ensemble in a manner best suited for its survival and refinement. It is the instrument of improvement, and without it animals would hardly rise above plants.' S. Ramon y Cajal (1911), *Histology of the Nervous System*, trans. N. Swanson and L. W. Swanson, Oxford University Press, New York.

There can be no doubt that the difference between the mind of the lowest man and that of the highest animal is immense...Nevertheless the difference...great as it is, certainly is one of degree and not of kind.

Charles Darwin (1871), *The Descent of Man,
and Selection in Relation to Sex*

A Japanese (female) monkey invented a way to separate food grains from sand (throwing handfuls into the sea); chimpanzees invented a method of collecting water by using moss as sponges. These are remarkable but exceptional events. Intelligent as they undoubtedly are, monkeys and apes don't invent much, though they may adapt rather wonderfully to living close to us (Figs 1 and 2). No other animal on earth has, or ever will, invent computers, husbandry on a massive scale, cars, houses, let alone write a poem or compose a symphony (although chimpanzees have produced paintings, it's not really clear that this involved truly artistic or aesthetic processes). Furthermore, the human brain endows us with the ability to ask questions about the natural world and about ourselves, and, through the invention of science, to supply at least partial answers, enabling progressive technical and social development. E. O. Wilson writes:

No matter how sophisticated our science and technology, advanced our culture, or powerful our robotic auxiliaries, *Homo sapiens* remains...a relatively unchanged biological species. Therein lies our strength, and our weakness. It is the nature of all biological species to multiply and expand heedlessly until the environment bites back. The bite consists of feedback loops—disease, famine, war and competition for scarce resources—which intensify until pressure on the environment is eased. Add to them the one feedback loop uniquely available to *Homo sapiens* that can damp all the rest: conscious restraint.

E. O. Wilson (2002), *The Future of Life*.
Little, Brown, London.

The human brain, together with an elaborate hand and a complex vocal apparatus, thus enables us to develop language, invent and make things and—equally important—develop complex and highly varied

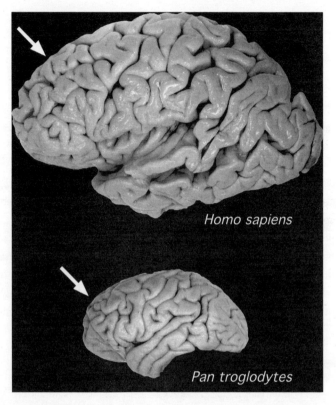

Fig. 1. The relative proportions of chimpanzee and human brains. Note the marked difference in the size of the frontal lobes (arrows) as well as the more elaborate patterns of folds in the human (indicating relatively more cortex). See Chapter 10 for more discussion of the human brain.

social structures. It also allows us to transmit to the next generation not only our genes, but through the invention of spoken and written languages the inventions, knowledge, and societal rules and traditions that may have been developed by previous generations. We don't fully understand the selection pressures that encouraged the enormous development of the human brain and thus these human attributes.

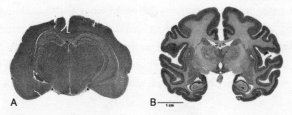

Fig. 2. Cross-sections (equalized for size) through (A) a rodent and (B) a primate brain. Note the increased complexity of the cortex in the primate, and its larger size relative to the rest of the brain compared to the rodent.

After all, other very successful species—rats are one obvious example, not to mention insects—do very well without needing a huge brain. Darwin himself was puzzled.

> More than one writer has asked, why have some animals had their mental powers more highly developed than others, as such development would be advantageous to all? Why have not apes acquired the intellectual powers of man? Various causes could be assigned; but they are conjectural, and their relative probability cannot be weighed.
>
> Charles Darwin (1872), The Origin of Species (sixth edition), edited by R. E. Leakey. Hill and Wang, New York.

We do know that the human brain had to develop its modern form before mankind began to alter his environment by building ever more elaborate shelters, get food by using ever more elaborate tools and weapons, and protect himself against cold by more elaborate clothing, and so on.[2] The rapid evolution of man since he developed his enormous brain has been cultural and technological, rather than physical. The latter took millennia: the former only a fraction of this timescale. But the important point is this: the human brain developed originally in response to the natural world, whereas the modern human brain shapes that world in a manner that promotes the well-being and survival of mankind. So while the composition of the natural, ancient world owes nothing to the human brain, the (human) modern world owes practically everything.[†] And here is the essential consequence: we bring to our modern world some of the properties of the brain that served us so well in that ancient one. But they have to operate in a very different environment from that for which they were originally developed and to which the human brain was first adapted.[3]

> We must, however, acknowledge, as it seems to me, that man with all his noble qualities, with sympathy which feels for the most debased, with benevolence which extends not only to other men but to the humblest living creature, with his god-like intellect which has penetrated into the movements and constitution of the solar system—with all these exalted powers—man still bears in his bodily frame the indelible stamp of his lowly origin.
> Charles Darwin (1871), *The Descent of Man, and Selection in Relation to Sex*

Reproduction is an obvious and prominent example. Successful reproduction is the endpoint of successful adaptation. 'Fitness', the hallmark of evolutionary success, is measured by an ability to transmit genes to subsequent generations: that is, successful reproduction. Reproduction is a complex process in all mammalian species, including ours. It involves fertility: the ability to produce viable sperm or eggs (gametes); mate selection: a competitive process that itself plays a role in the evolution of fitness; mating: to ensure that these gametes are fertilized; pregnancy: the development of the foetus; birth: production of live young; lactation; sustenance of those young; parental behaviour, to protect and nurture the newborn. Each part of this

---

[†] 'Long before the human species appeared, the pinnacle of evolution was already the brain . . . Animals with simple and primitive or no nervous systems have been champions at surviving, reproducing, and distributing themselves but they have limited behavioral repertoires. The essence of evolution is the production and replication of diversity—and more than anything else, diversity in behavior.' T Bullock (1984), quoted by E. M. Hull, R. L. Meisel, B. D. Sachs (2002), 'Male sexual behavior'. In: *Hormones, Brain and Behavior*, D. W. Pfaff, A. P. Arnold, A. M. Etgen, S. E. Farhbach, R. T. Rubin (eds). Academic Press, Amsterdam, pp. 3–135.

sequence carries risks and cost to both parents and young. One might imagine that, once this complex interlocking series of events had evolved to be a success, it would have become standard throughout the mammalian order.

But this is not the case. Although the objective of reproduction (biologists call this the 'ultimate' cause) is the same for all mammalian species, the way that it is accomplished (the 'proximate' cause) is astonishingly different. Reproduction is remarkable for being so varied between even mammalian species. Examine this more closely, and it is apparent that this variation lies mostly with the females (Fig. 3). So to understand the males' role in reproduction, we have first to consider the females', for it is she who sets the reproductive pace. Here are some examples. Female rats have an ovarian cycle that lasts 4–5 days, dominated by the production of oestrogen alone. They ovulate at this frequency then become sexually active (and attractive) for a few hours (in 'heat' or 'oestrous'). They produce progesterone, the other major ovarian hormone in small amounts (it's important for making female rats sexually receptive), but in much greater amounts if they mate. After a short gestation period, large numbers of very immature young are born, most of which will not survive. This is mass production and high infant risk. It serves rats very well indeed. Rabbits (and cats) also have a similar strategy, but it operates differently. These females can remain in heat (oestrous) for long periods, and only ovulate if they mate; in this way, they maximize the chances of becoming pregnant. Then the rest of the sequence (progesterone secretion) is activated, as in the rat, and they produce large litters, with a high risk of non-survival. Other species have different strategies. Species that produce many young, but are likely to lose many (i.e. high infant risk) are said to adopt an 'r' strategy. The alternative—higher investment in fewer offspring—is a 'k' strategy. Both have advantages and disadvantages, and they overlap somewhat. Sheep, for example, produce only one or two young, and have a rather long cycle with a correspondingly protracted period of sexual activity.

They don't need to mate or to become pregnant to secrete progesterone, which in their case is important for activating sexuality. They look after their young for comparatively long periods. Female monkeys and apes do likewise, though they don't need progesterone to enable sexual activity. Their cycle is rather like that of human females. There is thus a wide spectrum of variation between species in the way that females become fertile and mate. This carries on into pregnancy and parenthood.[4]

Many species, particularly those living in more temperate climates, time their births to occur in the spring, the season of increasing warmth and food supply. This requires a second tier of control: females restrict the costs of reproduction to only one part of the

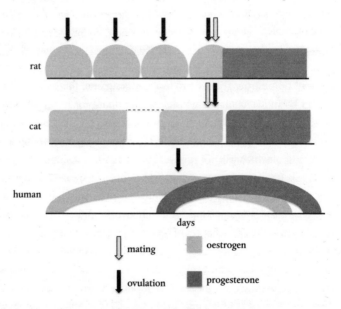

Fig. 3. Diagrams of the oestrous (reproductive) cycles of three female species. The rat has a 4–5-day oestrogen cycle, ovulates spontaneously, but only secretes much progesterone if she mates. The cat has a prolonged, and variable, period of oestrogen secretion: she only ovulates and secretes progesterone if she mates. The human female ovulates spontaneously after about 14 days of oestrogen secretion, and then has a similar period dominated by progesterone.

year, the breeding season. Females with a short pregnancy will become fertile early in the spring or late winter: those with a longer gestation (like sheep) will be fertile in the autumn. Some species, such as badgers and some deer, have evolved an even more elaborate timing mechanism: they can carry their fertilized foetus in a state of suspended animation in their womb, and only initiate its development in time for a spring birth. Even different species of non-human primates show highly distinct types of mating systems.[‡]

Human females have no obvious breeding season, though births are more common in the spring and autumn. They have a typical primate cycle, different from rodents and some other species. The first 14 days of each cycle are dominated by oestrogen secretion. Then the female ovulates (produces an egg) without the need to mate, and this is followed by a similar period of progesterone secretion, which prepares the womb for a future embryo. Humans, like other primates, produce one, occasionally more, well-developed young after a long gestation: a very different strategy from the rat and many other species. The growth of a comparatively large infant, particularly a large brain, requires a prolonged pregnancy, so that the newborn is better able to survive. But one consequence of this is a much bigger metabolic demand on primate mothers: the risk is to the mother rather than (or as well as) to the infant.[§] And her investment in each newborn is much greater than in those species adopting the rat-like strategy, because she produces so few. This will be reflected in the

[‡] 'Two important considerations [for defining primate mating systems] are, firstly, whether a female usually mates with one male, or more than one male, during the fertile phase of her ovarian cycle and, secondly, whether her sexual relationships are long-term and relatively exclusive or short-term and non-exclusive. This line of reasoning results in the recognition of five mating systems: 1. Monogamy, 2. Polygyny, 3. Polyandry, 4. Multimale-multifemale, 5. Dispersed or non-gregarious.' A. F. Dixson (1998), *Primate Sexuality*. Oxford University Press, Oxford.

[§] 'The daily energy budget of a nursing mother exceeds that of most men with even a moderately active lifestyle and is topped among women only by marathon runners in training.' Jared Diamond (1997), *Why is Sex Fun?* Weidenfeld & Nicolson, London.

increased care she gives her infant, and for how long she provides it.[5] It's a high maternal, low infant risk strategy.

Why is there so much variability in the way that female mammals reproduce? The obvious answer is that this relates, in some way, to differences in the environments in which each species has evolved or for which it has adapted. This would be the routine answer from any scientist interested in comparative biology. But if you were to press for a more precise answer—for example, what is it about a particular environment or evolutionary history that gives rise to these quite marked differences in reproductive methods?—you might get only a rather evasive reply. There is no clear general explanation, though for species that live in herds that are constantly on the move, the newborn must be able to keep up within a few hours of birth, and so are born comparatively mature. A second possible answer is that there is more than one way of achieving a goal, and different species have, for some reason, chosen different paths: an equally unsatisfactory explanation. It's a puzzle. Another possible reason for long pregnancies and a single young is that the development of the foetus requires it; for example, the growth of a large brain in man and other primates. But whereas rats give birth to a large litter of relatively undeveloped young (they look like foetuses at birth) many of which will not survive, guinea pigs—another smallish rodent-like species—produce fewer and much more mature young. The two species do live in very different habitats: guinea pigs (or, rather, their ancestors: modern guinea pigs are not found in the wild) in the tundra of South America, and rats live in all sorts of environments, some of them distinctly insalubrious.[6] But this does not explain their respective reproductive strategy.

Males of different mammalian species are much less variable. The principal species difference is in the timing of fertility, for the males of many mammalian species are also sensitive to the seasons, and only become fertile during part of the year. The exact time depends on the female. Since it is she who determines the season of birth, males have to time the onset of their fertility to coincide with hers. They dance to her tune. This is why a book on testosterone has to start with consideration of the way that females breed. The mechanism underlying the combined process of fertility and sexuality in males is rather similar across all species. The whole process depends upon activation of the testes. The testes produce the one essential signal that enables both parts of the process: testosterone. Testosterone has only one function—to enable a male to reproduce, for which it is essential. No matter how varied are the details of the reproductive anatomy, physiology, or behavioural strategies of the males of any mammalian species, there are basic similarities across the animal and human world. Testosterone is essential for all of them. Male reproduction, though not itself metabolically demanding, is socially and physically extremely perilous. So for reproduction to be successful, testosterone has to act on many parts of the male to make him fit for the competitive world of male sexuality. Simply enabling fertility and sexual motivation, essential as these are, is not enough. A male needs to compete with other males, and hence, if need be, behave aggressively and not be averse to taking risks—but knowing when the risk might be too great or when he is defeated. He also needs to be muscular, and maybe equipped with weapons (like antlers or long teeth) to enhance his competitiveness. Males need be sexually attractive, as well as competitive, for in most species females exercise their own selection over their sexual partners (Chapter 6 discusses this and related topics in more detail). Testosterone does all this, and more, which is the root of its importance and fascination.

A notable feature of sexual behaviour in the human male is the wide variety of forms it can take. This is in marked contrast to the patterns observed in other species, in which the expression of sexual activity is fairly stereotyped. The basic reproductive system in the human male is similar to that of other species, particularly other male primates. In this respect, humans conform to the general principle of consistency between males of different species, in contrast to females. So testosterone has a similar role in man as in other species. But many parts of the brain are not directly responsive to testosterone, yet play a vital part in its actions (see Chapter 10). The human brain, principally the cerebral cortex, endows men with the ability to devise complexity and variety to sexual behaviour itself (i.e. copulation), the context in which it occurs, and the social constructions around which sexuality is built, to a level which far exceeds that seen in any other species. In this way the human brain has taken an ancient and basic biological necessity and adapted it progressively to an evolving social and technical milieu.[7]

Sexual selection is a powerful influence on evolution. Both males and females judge the attractiveness of the other in a variety of ways, and because these qualities differ between individuals, this results in assortative mating. Assortative mating means that the direction of individual sexuality and breeding is not random. Humans have a huge array of criteria on which to determine their sexual selection. Some are based on individual choice (who finds whom attractive), but others depend on the social environment and historical epoch in which they live—the ways that the culture of a society determines, regulates, and influences the attractiveness or availability of particular individuals for others. Though the two influence each other, they are not necessarily the same. We will have more to say about this in subsequent chapters (e.g. Chapter 6). In the *Origin of Species*, Charles Darwin[8] wrote:

> This form of selection [sexual selection] depends not on the struggle for existence in relation to other organic beings or external conditions, but on a struggle between the individuals of one sex, generally the males, for the possession of the other sex. The result is not death to the unsuccessful competitor, but few or no offspring. Sexual selection is, therefore, less rigorous than natural selection. Generally, the most vigorous males, those which are best fitted for their places in nature, will leave most progeny. But in many cases victory depends not so much on general vigour as on having special weapons confined to the male sex.

This essential fact has even seeped into literature:

> The Greek, the Turk, the Chinese, the Copt, the Hottentot, said Stephen,
> all admire a different type of female beauty. That seems to be a maze out
> of which we cannot escape. I see, however, two ways out. One is this
> hypothesis: that every physical quality admired by men in women is in
> direct connexion with the manifold functions of women for the propaga-
> tion of the species.
>
> James Joyce (1916), *A Portrait of the Artist as a Young Man*

Testosterone plays a central and irreplaceable role in sexual selection, since it has powerful influences both on a male's sexual motivation, and upon his appearance and behaviour towards potential partners.

Sexual selection conceals an intriguing paradox. It is assumed, for good reason, that females—who make the final choice in many species—tend to choose males that display attributes that declare themselves to be particularly 'fit': that is, individuals who are success-ful competitors, able to protect them and produce offspring that are themselves highly competitive and successful in the frantic world of sexual competition. But this will apply only to their male offspring. For a female, 'fitness' may imply a different set of characteristics, those more associated with fertility or a good standard of maternal care. These are not necessarily those displayed by the male. So how does a female select a male that might be more likely to provide her with female offspring with these alternative characteristics? Males, of course, can—consciously or otherwise—select females that have fea-tures indicating good potential for breeding. So maybe there is a reciprocal process during courtship, with each sex selecting the future properties of the opposite one. But if females really do have the more prominent role in sexual selection, then the process seems biased towards the fitness of males (see Chapter 6).

Successful human males in their primeval world presumably had all these testosterone-dependent qualities, which they share with other male mammals; how do they function in the modern world? Have they survived essentially intact, or has the same modifying process that man has applied to other parts of his world also been applied to mating, selection and reproduction? Does testosterone do a similar job today as it has always done, or has its role been fundamentally changed in the complex and changing circumstances of modern life? Does it have other roles today, not imaginable in earlier, simpler cultures? Has testosterone shaped any part of our world? How has this world dealt with the age-old problem of testosterone—how to contain its powerful actions so as to be compatible with a stable and successful society, whatever the outward form and customs of that society? These are some of the questions that are debated in the rest of this book.

We will see that, even in the modern human world we like to think of as sophisticated and complex, testosterone continues to play its ancient role. But to do this, it needs to exert an influence on almost every aspect of our lives. The tendrils of its powerful actions creep into much of what we do. By 'we' I do not simply mean men: the dramatic, but often unrecognized, influence it has on men is reflected in women, since the way men behave has such an impact on the lives of women. Testosterone has important roles in women, too, as we will see, though these are less often recognized than those in men (Chapter 9), and thus on another dimension of the formation and structure of the world we all live in. It has shaped the kind of people we are, the things we invent or prefer, and the kind of society we live in, determines how men relate to each other, to women, and how groups of men interact with other groups, in ways we are beginning to understand. And it goes on doing so, in the world in which we now live. Can we really ascribe all this to one simple chemical? This is what we explore in this book: the molecule that has made our history.

Religion, psychoanalysis, philosophy, ethics...all have thought of humans beings as a mix of 'light' and 'dark', 'superego' and 'id', 'good and evil', and so on. This is not the view taken here. Primeval or ancient patterns of behaviour are neither 'good' nor 'bad'; they are certainly not evil. They have evolved as ways of coping with demanding, difficult and varying environments. More recent patterns or control of such behaviours may have moral values to those concerned with such topics, but here we think of them simply as examples of the way the extraordinary human brain has adapted over the centuries to changes in lifestyle and circumstance. We can draw all sorts of conclusions from the way we have adapted, and the differences across time and culture in the way this has happened. No doubt the direction of these changes, which are not preordained but subject to the flexibility which so characterizes the human brain, call on all these other aspects of the human condition, and the human ability to devise or apply laws, ethics, traditions, and moral standards to fundamental patterns of behaviour. It may not be the substance that enables humans to be so inventive, but it adds essential impetus to achieving that inventiveness. So when we consider how the brain responds to testosterone, do not expect an account of a superior, highly developed 'human' part of the brain holding a more demonic region in check. Despite the enormous changes in the human condition over the centuries, and all the different shapes these have taken, the fundamental fact remains: reproduction is the essential function of all species, including man, and testosterone continues to play a central role in it. But the ability of the human brain to devise increasingly elaborate ways of moderating and extending that role is what this book is all about.

One word of caution. When we try to recapitulate our biological and social history, we are limited by the records available to us. Mostly, they go back only a few thousand years, if that; and they are often incomplete, or may not contain the information we seek. So we are tempted to look at other 'lesser' species—particularly other primates—for hints on what that history may have been. But we need constantly to remind ourselves that these species, too, have been evolving and adapting for even longer than us. During that time, they will have departed to a considerable extent from an earlier design. So they are an uncertain guide to our own history, though there may be clues—common features that persist. Rats, cats, dogs, monkeys, apes, and man: this is most certainly not a straightforward evolutionary sequence, but represents complex branches off a common, very ancient, trunk. The problem is to distinguish features that represent a common ancestry from those that result from subsequent evolutionary adaptations. The important differences between comparative studies and those on evolution are well understood, but still frequently confused. Even Darwin could be said to have crossed this line occasionally! It would be prudent to recall this caveat as you read this book. But there is no getting away from the fact that to understand our present, we need to enquire into our past (Fig. 4).

> Evolution is central to the understanding of life including human life. Like all living things, we are the outcome of natural selection; we got here because we inherited traits that allowed our ancestors to survive, find mates, and reproduce. This momentous fact explains our deepest strivings: why having a thankless child is sharper than a serpent's tooth, why it is a truth universally acknowledged that a single man in possession of a good fortune must be in want of a wife, why we do not go gentle into that good night but rage, rage against the dying of the light.
>
> Steven Pinker (2002), *The Blank Slate*. Allen Lane, London.

A second word of caution. It is all too easy to substitute 'masculinity' with 'testosterone' or even ascribe all gender differences, physical, social, and political, to its action. Testosterone certainly has a powerful impact on the lives of men. But, as already mentioned, testosterone also has important roles to play in women, more fully discussed in Chapter 9. There are even those who would ascribe anything they find unpleasant about male behaviour as being 'just testosterone'. How far are these attitudes justified? These are also some of the issues with

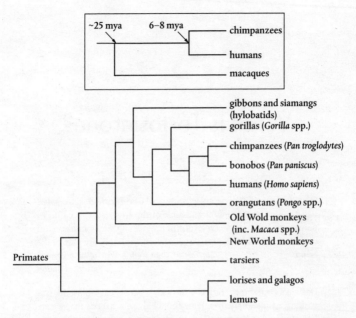

Fig. 4. The evolutionary tree of the primates. Chimpanzees are man's closest relative, but even they diverged from the human line 6–8 million years ago, and have pursued an independent evolution ever since.

which this book will be concerned. Testosterone is not a metaphor for a man; but it is a powerful hormone, whose influence it is easy either to ignore or to overestimate. And its effects on our behaviour and lifestyle are all around us. But first, let's consider what we know about testosterone itself.

2

What is Testosterone?

O appendage of maleness! Lightning rod of misfortune, antenna of bane, tower of reproduction, trunk of the tree of life, column of perpetuity, pillar of the race, tube of the rushing vital wind, conduit of excretion and hose-pipe of generation: you, too, are part of the body.　　　　F. Gonzalez-Crussi (1986), *Notes of an Anatomist*.
Picador, London

Darwin's contemporaries saw at once what a heavy blow he was striking against piety. His theory entailed the inference that we are here today not because God reciprocates our love, forgives our sins, and attends to our entreaties but because each of our oceanic and terrestrial foremothers was lucky enough to elude its predators long enough to reproduce.

F. C. Crews, *New York Review of Books*, October 2001

The body is made up of numerous organs, and numerous cells within those organs. To function as a whole and to adapt as a whole to whatever circumstances may arise requires communication between cells and organs. There are several ways in which this can happen, but hormones are an important and essential channel by which information is transferred around the body. So it is that levels of blood glucose and the way this is taken up to be used as fuel by cells is controlled by several hormones, of which insulin is the most prominent. Your blood pressure is controlled by another set of hormones, and the constituents of the blood such as calcium (essential for muscles to work) by yet more hormones. Hormones enable you to cope with stress, since they alert the whole body for action. Some hormones come from specialized glands, like the thyroid or adrenal, but in recent years it's been found that hormones are also secreted by tissues that had never been thought of as hormone-producing or 'endocrine' glands. So even the heart, usually considered a pump, produces hormones that help regulate fluid in the body (and hence blood volume), and fat, for long thought to be merely a storage site for excess energy, is now known to secrete a variety of hormones, one (leptin) that regulates long-term body weight.

Just below the brain, directly behind your eyes, is the pituitary, a gland that secretes a variety of hormones.* Some have a direct action on the body: for example, growth hormone, which regulates growth in children, but has other actions (on metabolism) in adults. But other pituitary hormones do not act on the body as a whole, but on other hormone-producing glands (e.g. the thyroid and adrenal). Two of these hormones, called 'gonadotrophins',[9] regulate how much testosterone is produced by the testis (Fig 5A&B). Pituitary gonadotrophins, in turn, are controlled by the brain. Hormones are thus chemical messengers, produced by a local collection of specialized cells (a gland) and secreted into the bloodstream.

Testosterone is also a hormone. Nearly all testosterone in men comes from the testis (a tiny amount comes from the adrenals). It's an extremely ancient hormone; it seems to have been developed early in the evolution of vertebrates. Birds, fish, reptiles, and mammals all produce testosterone, and rely upon it for reproductive competence. Testosterone has evidently been a huge evolutionary success story: despite the enormous differences between these groups of animals,

---

* Sometimes called 'the conductor of the endocrine orchestra', or 'the master gland'.

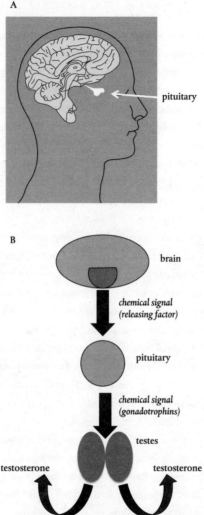

Fig. 5. (A) The pituitary, which controls the activities of the testes, lies at the base of the brain. (B) The control of testosterone secretion. Neurons in the hypothalamus (at the base of the brain) produce a chemical signal (a releasing factor) that acts on the pituitary; this, in turn, secretes large peptide hormones (gonadotrophins) that act on the testes. Gonadotrophins enable both the formation of sperm and the production and secretion of testosterone into the blood. There is thus a chain of command, starting in the brain, though this can be moderated by levels of testosterone in the blood.

and even within the groups themselves, testosterone has remained constant and essential for them all.

A word about the testis. It's a truly remarkable little gland (it weighs around 20 grams in man). It contains two quite distinct mechanisms that together are ultimately responsible for every man's fertility and sexuality, and hence for the continuation of humankind (as well as all the other vertebrates). The fertility function depends on a system of numerous tubules, each lined with cells that make sperm—itself a remarkable achievement, since it involves halving the number of chromosomes each sperm-germinator cell (spermatozoa) contains, accurately and consistently (in man, from 46 to 23: very occasionally it goes wrong). The normal number will be restored when the sperm fuses with an egg, itself containing only half the usual number of chromosomes. This tubular system in the testis ensures that the newly formed sperm are delivered to the ducts that carry them to the penis, and hence to the waiting vagina. Scattered between these tubules are clumps of cells with quite a different function: they make and secrete testosterone (Fig. 6).[10] They have no need for tubules, for testosterone goes straight into the blood, though it also passes into the tubules,

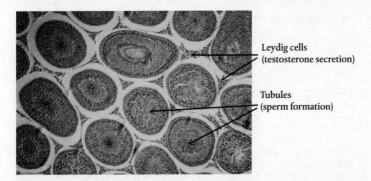

Leydig cells
(testosterone secretion)

Tubules
(sperm formation)

Fig. 6. A microscopic section through the testis to show the system of tubules (making sperm) with clumps of hormone-producing Leydig cells in between them.

21

where it has an important role to play in sperm formation. Though most men have two testes, they actually only need one.[11] The testis is the source of most of what we term masculinity; what this means, and how this is achieved, is what this book is about.

The blood acts as a transport system, and since it pervades the whole body, hormones are able to reach every corner. That doesn't mean that every cell responds to every hormone. To respond to a hormone, a cell needs another complex molecule: a receptor. Organs that have such receptors are called 'target' tissues.[12] A receptor not only detects the presence of the hormone, it also starts the process whereby the cell responds to the hormone. Exactly how it responds depends on the hormone and the cell: each hormone can act on a variety of cells in different organs, and each will respond differently (Fig. 7A). Some hormones have more than one receptor, so whether a cell possesses one or the other, or even more than one, will also influence how it responds. The patterns of receptors throughout the body—and these can change—thus determine how the body responds to each hormone. This also means that the body has at least two ways of regulating hormone action: altering the amount of hormone secreted by the gland, but also the amount or sensitivity of the hormonal receptors. So receptors are as important as hormones themselves.[†]

You can see that measuring levels of hormones in the blood, a common way of assessing their activity, gives only partial information. Imagine trying to assess the use of a particular commodity by measuring the number of lorries on a motorway carrying this product.

[†] 'One of the major determinants of a transmitter's action is the receptor mediating the target's response. This idea is so simple and self-evident that it is easy to forget that in the early days of chemical transmission research no one knew about receptors, so the possibility of receptor diversity was not a formal concept...As it became clear that a single transmitter could exert very different responses with different latencies and durations, the notion of receptors came into being.' B. A. Trimmer (1999), 'The messenger is not the message; or is it?' In: *Beyond Neurotransmission*, P. S. Katz (ed.). Oxford University Press, pp. 29–82.

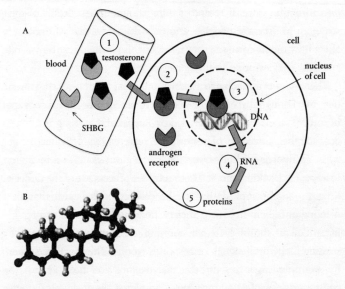

Fig. 7. (A) The process by which testosterone acts on cells. (1) Testosterone in the blood is either bound to the protein SHBG or floats free. The free fraction is able to enter the cell where (2) it binds to its receptor protein. (3) This complex then moves into the nucleus and binds to DNA. (4) DNA then makes a special set of RNA molecules that, in turn, (5) enable the formation of particular proteins. (B) The testosterone molecule. The dark spheres are carbon atoms, the pale ones are hydrogen, and the two grey ones are oxygen. Note that the backbone of testosterone is made up of four rings of carbon atoms. Three of them are formed from six interlinked carbon atoms, and the fourth (on the right in this diagram) has five. This is what constitutes a steroid.

Of course, this does give you some idea of how much is being produced (though you might also want to know how much goes into the factory's warehouse for storage: some glands also store hormones as well as secrete them). But we would lack information about where the lorries are going or who uses the product. Simply measuring how much hormone there is in the blood will not tell us what happens in each tissue—whether there are receptors and, if so, what they do; whether the hormone is used as it is, or metabolized into something else. If receptors differ in amount or details of their structure in different people—and they do—then this will not only affect how much hormone they need, but also how much they secrete and how they respond to the hormone itself.

There has been much discussion about what constitutes a hormone (Fig. 7B). Though there have been numerous claims to have discovered hormones, the term was first used by Bayliss and Starling in 1902 to describe their discovery of a substance, secretin, that activated the gut without involvement of nerves. Bayliss received in 1922 an invitation to appear at Buckingham Palace to receive the accolade of the Order of Knighthood; he replied that as the date coincided with that of a meeting of the Physiological Society, he would be unable to attend.[13] Every cell in the body either communicates with another cell by sending a chemical signal, or responds to one. The immune system, for example, organizes the essential defence you need against the constant risk of infection by using a complex series of chemical signals that activate, de-activate or transform the myriad cells that make antibodies or destroy invading bacteria. Your gut processes the food you eat by responding to signals released by the liver, gallbladder, and other parts of the gut (it also produces hormones that alter appetite). The brain is an electrochemical machine. Nerve cells (neurons) are connected together in unbelievably complex networks: but each cell communicates with the next using a chemical signal—a neurotransmitter. Are these all hormones?

A great deal of time has been wasted on such debates. If one limits the term 'hormone' to a chemical messenger released into the blood (the original definition, one made before other types of cell–cell signalling were known) then one is faced with the problem that the same chemical can act both as a hormone and a shorter-range chemical signal. One example is noradrenaline (called norepinephrine in North America). Released into the blood by an acute stress or excitement, it causes your heart to pound and your blood pressure to rise. But the same molecule is released in the brain (under rather similar circumstances), where it acts as a neurotransmitter on other neurons only a few microns (1 micron = 1/1000 mm) away without entering the blood. Cells in the immune system act on other cells slightly further away (some via the blood). Most people stick to the original definition, but a more sophisticated view is that chemical signalling between cells can take place in various ways and over various distances: the principles are the same. These signals contain a wide range of information, and enable the body to work in a coordinated and efficient manner. The classic hormone is one example.

Testosterone shows all the features of a classic hormone. As already mentioned, in the male nearly all testosterone is secreted by the testes into the blood, but a tiny amount comes from the adrenals, glands that lie on top of the kidneys (hence their name). The adrenals are really concerned with secreting another set of hormones, including the 'stress' hormone cortisol, as well as other hormones concerned with salt balance and (maybe) ageing. In women, some testosterone comes from the ovaries and, because the overall levels are so much smaller, the amount secreted by the adrenals may be more important (this is discussed more fully in Chapter 9). These hormones, together with testosterone and, in females, oestrogen and progesterone, are all steroids.

A steroid is a chemical derived from cholesterol, a substance more familiar as something to watch over, and worry about, in the blood. But cholesterol has many important functions, and one is to provide the starting point for the body's manufacture of steroid hormones. The testes have a series of enzymes,[‡] and these progressively shape the cholesterol molecule by chopping bits off it, or adding atoms to it, rather like a sculptor modelling a block of stone into a finished figure,

---

[‡] An enzyme is a complex protein that is able to change the structure of another substance (often itself a protein). They are responsible for the entire process of metabolism (and many other functions).

in this case testosterone. Other glands use different enzymes to produce different steroids, as another sculptor might make a different figure from the same stone in another studio. Just as such sculptures may have a generic similarity, so steroids resemble each other because of their common origin. They can have very different actions (e.g. testosterone vs. oestrogen), because small changes in their molecular structure can have profound effects on the way they act.

Since each steroid has widespread actions on the body, in a clinical setting it can be advantageous to select only those that are required; the others will be unwanted side-effects. This has proved a fertile and profitable field for biochemists and pharmaceutical companies, who have, very successfully, devoted vast resources to produce artificial steroids that can imitate, block, or reproduce some selected parts of the overall action of the natural one. These compounds don't always look very like a natural steroid: this is because the receptors can be fooled into thinking a molecule resembles the natural steroid if it fits onto the receptor in certain ways. A steroid called stilboestrol was one of the first molecules to be made: it mimicked the effects of oestrogen when given by mouth (oestrogens are not very effective by this route) and was used for many years to treat prostate cancer. Tamoxifen, a steroid that blocks the action of oestrogen, is often used to treat those cancers of the breast that are sensitive to oestrogen. A major industry has been to make artificial types of cortisol, the stress hormone. Cortisol, as part of its many actions, reduces inflammation and allergic or immune responses. But it has other effects, which may not be so welcome. So these artificial steroids can replicate some of the actions of cortisol, though not the less desirable ones. This is because the receptors for cortisol are not the same in all tissues. One of the ways of doing this is to make a compound that is highly selective for one type of receptor. Cortisol acts on two receptors (it may have other actions as well), so limiting the effect of an artificial steroid-like drug to only one will reduce unwanted side-effects.

The sculpturing of steroids doesn't end when they are secreted. The target tissues can carry on the process. Testosterone, for example, is converted by another enzyme into DHT (dihydrotesterone) by target tissues such as the prostate gland. Since DHT is a more potent male hormone (androgen) than testosterone itself, blocking the enzyme that makes it can be helpful in some clinical situations. An artificial steroid called finasteride inhibits conversion of testosterone to DHT; since some cancers of the prostate are very sensitive to DHT, preventing its formation reduces the cancer's growth. Even more intriguingly, other tissues (including the brain) convert testosterone to oestrogen, which raises serious questions about the validity of talking about 'male' (androgens') and 'female' hormones (oestrogens). But giving oestrogen has very different effects to those of testosterone, so many of the actions of testosterone do not depend on its conversion in target tissues to oestrogen.

There was a time when people began thinking of testosterone as a 'pro-hormone'; that is, a steroid which was converted by target tissues into more potent steroids that actually were responsible for their effects. It has now become clear that there is a receptor (the androgen receptor) that responds to testosterone directly, though DHT has a more potent action on this receptor than testosterone. Oestrogens, on the other hand, have a separate set of receptors, so if cells are to respond to oestrogen derived from testosterone they need these receptors as well as the androgen one.

Let's return, for a moment, to our analogy about lorries and products. When testosterone enters the blood from the testis, it encounters a very large protein called SHBG (sex hormone-binding globulin). As its name implies, testosterone binds to this protein, rather as if a carton of goods had been loaded onto a lorry. Attached to this protein, testosterone is whisked around the body by the impetus of the blood flow. Binding to SHBG may protect testosterone from being degraded by other proteins, and prolong its life. But not all testosterone jumps on this lorry: some (only about 5%) remains floating free in the blood, along-side the 95% bound to SHBG. The difference is highly important because only the unbound testosterone is available to the target tissues (this includes the brain); they can't access the 95% bound to SHBG. Why, you ask, is it there? What is its use? The answer is that the bound testosterone can detach from SHBG, and does so when the free testosterone falls below around 5% (we could say it falls off the back of the lorry). So SHBG-bound testosterone acts as a reservoir for more free testosterone when it is needed. There is a constant balance between the two forms of testosterone. This means that measuring all the testosterone in the blood, which we have already seen doesn't give all the information one might need, is even less useful (though by no means useless) than we had thought. We need a measure of the 'free' hormone. There are various ways of doing this, but measuring levels in the saliva is a convenient and relatively simple one. Since only 'free' testosterone can enter the tissues (in this case the salivary gland), what passes into the saliva represents only 'free' hormone.

The brain is a very delicate organ, and it needs a very special and protected niche. Surrounded by solid bone that safeguards it against changes in the physical environment, it is also surrounded by the blood–brain barrier, a structure formed largely by its blood vessels that protects it from rapid changes in the composition of the blood. Fluctuations in the composition of your blood (e.g. after meals) don't necessarily upset the delicate environment in which the brain sits. Some molecules, particularly big ones like proteins, can't get into the brain at all, unless the blood–brain barrier is damaged in some way. The important point for us is that only 'free' (unbound) testosterone can enter the brain. Testosterone bound to SHBG makes a complex that is far too large to get in. So if we want to get a realistic idea of how much testosterone is reaching the brain, we need to measure free testosterone in the blood. A direct way of knowing how much there is in the brain is to examine testosterone in the fluid surrounding the brain—the cerebrospinal fluid (CSF). Any substance in this fluid has unfettered access to the brain itself. But CSF is hard to get at. You need to put a needle into the base of the spine (and even this may not tell you everything about what is going on in the brain), or stick one into the back of the neck (only for very skilled people!), or drill a hole in the skull (ditto). Not everyday procedures, and certainly not ones that could be justified as part of research (though they are all used clinically from time to time). So saliva is a good substitute, and has been used for this purpose in humans. A corollary of this is that anything that alters SHBG levels (and hence 'free' testosterone) will, among other things, alter the amount getting into the brain, even if total levels in the blood do not alter.

SHBG, like our lorries, can only carry so much testosterone. As testosterone levels rise, therefore, there will come a point at which the carrying capacity of SHBG is filled. After this, any further testosterone will simply stay in the 'free' form, and thus be available to enter the brain as a surge. So, again, if we measure total testosterone in the blood without any knowledge of SHBG, we may miss this tipping point, and underestimate the surge of testosterone that is entering the brain as levels rise above the capacity of SHBG. This means that the relation between blood and brain levels of testosterone is not always simple or linear, a fact that may be relevant for those who take testosterone.

So testosterone, or rather 'free' testosterone, enters the tissues. What happens then? (Fig. 7A) Cells are contained within a bag-like membrane, rather like the skin of a balloon. Most (non-steroid) hormones can't penetrate this membrane, but have to settle for reacting with receptors that lie in the membrane itself. What happens next is outside their immediate control. Steroids are different. They pass effortlessly through the cell membrane and enter the complex interior of the cell, full of different organelles and chemicals, the cytoplasm. It is here that they encounter their receptors, if they are present. Not all cells make these receptors; in particular, those for testosterone (the androgen receptor) are found only in tissues that respond to it. For example, muscles, the penis, and the prostate are all tissues that 'express' (i.e. make) the androgen receptor. At puberty, therefore, a boy develops bigger muscles, his penis enlarges, and (unknown to him) so does his prostate (it may become a problem later). The brain also responds. But not all of the brain has androgen receptors. We will consider the role of testosterone in the brain in more detail later; here, we will just note that those parts of the brain responsible for motivation and emotion are particularly rich in androgen receptors, a fact that agrees very nicely with what we know about the effect of testosterone on behaviour.

The androgen receptor, like other receptors, is a large complex protein, lying in the cytoplasm (see Fig. 10A). It has several parts: one 'binds' to testosterone: that is, testosterone becomes attached to it. Scientists like to talk of 'locks' and 'keys', implying that the lock (the receptor) is designed in such a way that only the right key (testosterone or DHT) can bind to it. This chemical binding has dramatic effects. Astonishingly, the receptor, with its bound testosterone aboard, now makes its way through the cytoplasm to the cell's nucleus, the control system of the cell. It only does this when it has testosterone bound to it. Getting into the nucleus is not so easy as passing through the cell membrane, but special mechanisms, sensitive to the 'bound' receptor, make it possible. Testosterone acts as a kind of password. After entering the nucleus, a second astonishing event occurs. The receptor, again enabled by its bound steroid, seeks out a special length of the DNA code and attaches itself to it. This strip of DNA, the 'steroid response element' is actually part of a gene or, rather, a great number of genes. Attaching the bound receptor to the strip of DNA activates or represses the gene (depending on which gene it is). Since genes in a cell regulate its function, this will alter the activity of the whole cell and thus the target tissue (Fig. 7A). If it is a muscle cell, it will increase its size, and the individual will appear more 'muscular'. Prostate cells will start to divide, so the prostate will enlarge. The hair follicles on the face will start to grow hair and a beard sprouts. Those on the forehead will stop making hair, and the typical male hairline recession appears. The voice deepens (there are reports that basses have higher testosterone levels than tenors[14]). Similar events occur on other parts of the body, but only those that possess the essential androgen receptor. This direct control of genes by a steroid hormone is remarkable; most signals, as we have seen, stop at the cell's surface. Incidentally, viruses can also access your DNA directly: in fact, they have to do so in order to replicate and, in some cases, make you ill. But not much else does—except steroid hormones.

Testosterone is the same molecule in all men and, indeed, all vertebrates. But the androgen receptor isn't. The gene that makes the androgen receptor, interestingly, is on the X chromosome. Males have only one X chromosome, so its function is very important, but females have two (and no Y chromosome—see Chapter 3). The fact that Tom, Dick, and Harry look, sound, and behave differently may have a lot to do with their different upbringing or life-styles, but more to do with the fact that, although the three all have the same genes, these genes are not identical. There are several million possible variations in genes, which accounts for the recurrent astonishment of parents that their children, growing up in much the same environment, can turn out to be so different.[§] Life-styles, of course, can also be influenced by genes, so there is a reciprocal interaction between genes and environment. Like all other genes, the one for the androgen receptor varies between different men.

There are several mechanisms that, together, cause genes to vary so much, but the one that applies to the androgen receptor is a CAG

[§] 'Personal uniqueness itself says something useful: molecular biology has made individuals of us all. Genetics disproves Plato's myth of the absolute, that there exists one ideal form of human being from which there are rare deviations such as those who have an inborn disease.' S. Jones (1993), *The Language of the Genes*. HarperCollins, London.

repeat. DNA is made up of a long chain of four molecules, like four different kinds of beads on a very long necklace. The four, A (adenine), T (thymidine), C (cytosine), and G (guanine) can occur in any order, but the order is important. The genetic code consists of a long stream of triplets (e.g. ATT, CGA, etc.) that are translated, through RNA, into proteins. Proteins are made up of a long string of amino acids, and each DNA triplet codes for one amino acid. The sequence of amino acids determines what the protein will do and how it works, and this, ultimately, depends on the genetic code. Some of the DNA triplets can be repeated, and the number of such repeats can be very significant indeed. Near one end of the androgen receptor gene there is a CAG sequence that can be repeated from about 10 to about 30 times in different men. The important point is that the sensitivity of the androgen receptor to testosterone is inversely related to the number of repeats: so those with longer numbers are less sensitive to testosterone. This may also affect testosterone levels, since the sensitivity of the androgen receptor is part of the control system regulating the secretion of testosterone from the testes. If the CAG repeat number goes above 30 (which is very rare) then there is a surprising, and tragic, consequence. Kennedy's disease is an inherited debilitating neurodegenerative disease resulting in muscle cramps and progressive weakness due to degeneration of neurons in the brain stem and spinal cord; it is caused by this longer repeat, though we don't really understand why. Incidentally, this is not the only neural disorder caused by having too many CAG repeats. Another example is Huntington's disease, a rare genetic disorder resulting in involuntary movements and progressive dementia, which is the result of having too many CAG repeats in another gene (called, naturally, 'huntington').

If we needed an even more dramatic example of the central importance of the androgen receptor then here is one. There are rare mutations of the androgen receptor that make it unable to bind to testosterone or DHT, and thus enter the nucleus and influence

DNA. If the failure to bind is complete (this will depend on the exact nature of the mutation), then the individual, though genetically a male with a Y chromosome, testes, and testosterone just like any other male, is completely unaware of his own testosterone and will grow up to look like, feel like, and be like a female—though one without a uterus or ovaries. The condition is called the 'androgen insensitivity syndrome' (AIS) and is a glaring demonstration that testosterone is at the root of what we call 'masculinity'.[15] We will consider AIS further in a subsequent chapter. All this emphasizes the important—but often underestimated—role of the androgen receptor and its variants on the way that testosterone functions. Only recently have technical advances made detailed analysis of this receptor a practical proposition in clinical or research settings. Interestingly, there exists a mutation of the androgen receptor in rats, rather similar to that in humans, which results in much the same picture: the 'male' rat, though complete with a Y chromosome, looks like and behaves like a female (see Chapter 3).[16] This and related matters are discussed in more detail in the next chapter. There is much more to be discovered about the androgen receptor. It may turn out to play an as-yet underrated role in the way that men vary, not only in their sexuality, but in many of the other traits associated with masculinity.

So testosterone does three important things for an adult male: it enables him to be fertile; it causes him to grow features, such as beards, hair, and muscles, that both enhance his sexual attractiveness but also ready him for the competitive and risky life of an adult; and it acts on his brain, not only to make him interested in sex and seek it out, but give him the psychological and emotional qualities that help make that venture successful.

Though testosterone may be a constant feature of all men, the way that each man responds to his testosterone can differ quite markedly. Variations in the androgen receptor is only one way that this comes about. Although we focus on testosterone, we shall need to keep reminding ourselves that what it does to a man is moderated by many other factors, including other genes. All men may have testosterone, but they are not all the same. One obvious example is baldness: testosterone induces baldness, but not in all men—though it causes a characteristic recession of the hairline over each brow in most. But whether it occurs, and to what extent, depends not only on testosterone but also on many other factors, including particular genes (baldness is strongly heritable). It is not difficult to imagine that many of the other actions of testosterone can vary on different parts of the body, or even parts of the brain, between different individuals. In discussing generalities, it is important not to lose sight of individuality, the basis for selection and survival. But the roles of testosterone are not limited to the adult. Testosterone has a powerful role in the development and shaping of a male.

3

Testosterone Makyth Man

From the earlier work...we concluded that the fetal period is critical for the psychosexual masculinization of the nervous system. However, further work with the rat convinced us that it is not the fetal period per se that is critical but rather the period of differentiation, whether prenatal or postnatal. Moreover, not all periods of fetal or neonatal development are equally susceptible to the masculinization action of testosterone.

C. H. Phoenix (1974), 'Prenatal testosterone in the nonhuman primate and its consequences for behavior'. In: *Sex Differences in Behavior*, R. C. Friedman, R. M. Richart, R. L. Vande Wiele (eds). John Wiley, New York

Welcome O life! I go to encounter for the millionth time the reality of experience and to forge in the smithy of my soul the uncreated conscience of my race.

James Joyce (1916), *A Portrait of the Artist as a Young Man*

Long before it's about hairy chests and deep voices, testosterone has made its mark. Every cell in your body contains 23 pairs of chromosomes, each a strip of DNA encoding thousands of genes. All the pairs differ somewhat from each other but seem to be made up of two rather similar chromosomes. Except one pair: in a male one of the pair is quite large, and looks rather like the rest of the pack. But the other is tiny: you might be forgiven for overlooking it. You would be wrong. This is the Y chromosome, and its presence makes a male. The other, larger, one is the X chromosome; males have only one, in contrast to the two possessed by females.

The number of chromosomes is halved during the formation of an egg, each egg receiving one of a pair. So all of the female's eggs have only one X chromosome. In the male, about half of his sperm have an X chromosome, but the others have a Y. If an X sperm fertilizes an egg, the result is a female: the two sets of chromosome combine to restore the original 46, so the new embryo now has XX (female). But if it's a Y sperm, then the embryo now has XY: and that's the recipe for a male (Fig. 8). Very occasionally, babies are born with additional X chromosomes together with a Y (e.g. XXY, XXXY, etc.). Remarkably, despite any number of (large) X chromosomes, the presence of even one (tiny)

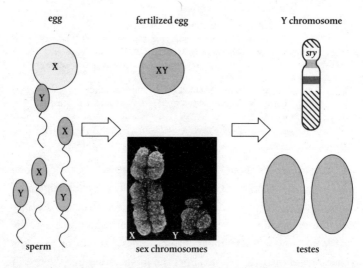

egg   fertilized egg   Y chromosome

sperm   sex chromosomes   testes

Fig. 8. How a male is made. If a Y-carrying sperm fertilizes the egg, then it forms an XY embryo. The presence of a Y chromosome (shown together with the X in the insert) and its associated *sry* gene enables the formation of the testes.

Y ensures that they are male.* The Y chromosome, being so small, has only a few genes on it (about 70, other chromosomes have about 2,000). But one, the *sry* gene, really matters. Together with help from some other genes, it does something really important: it makes a testis (two, actually). Without a Y chromosome (or no *sry* gene), the embryo doesn't make a testis, and becomes a female. Interestingly, the X chromosome has several genes that enable the testis to make sperm—a kind of genetic collaboration.

The advent of a testis is a hugely significant event. The new testes quite promptly start making and secreting testosterone (during the first couple of months of a human embryo's life). This testosterone has major effects on the internal structure of the new embryo. It enables formation of the reproductive structures that are essential for a male to become fertile (e.g. the ducts that carry the sperm from the testes to the penis). It makes a penis. It also sensitizes various tissues, such as muscle, to the effects of testosterone later in life. The embryo's brain is developing fast: the newly arriving testosterone acts on this as well, in ways that we'll consider later. But they are powerful and permanent actions that last for a lifetime. Then, just as suddenly as they started, the testes almost stop secreting testosterone.[17] So testosterone levels in a boy foetus during the last half of a human pregnancy are much lower. But the job's done. The little embryo is finally and permanently a male. Testosterone has made its first mark on its fate. Not only has it predestined its future as a male, it has also ensured that the body of this little male will be highly sensitive to later testosterone. Everything is set for a

---

* Individuals with XXY are known as Klinefelter's syndrome. They develop fairly normally, but may have reduced levels of testosterone and somewhat impaired ability to learn language. It's not that uncommon (about 1 in every 650 births). Increasing numbers of X chromosomes (e.g. XXXY, XXXXY: much more rare) result in greater cognitive defects. See A. Gropman and C. A. Samago-Sprouse (2013), 'Neurocognitive variance and neurological underpinnings of the X and Y chromosomal variations'. *American Journal of Medical Genetics*, vol. 163C, pp. 35–43.

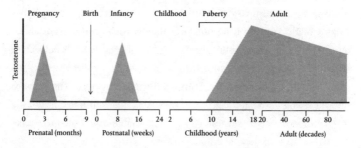

Fig. 9. The three surges of testosterone secretion during a man's life. Episodes occur during foetal life, again shortly after birth before the more sustained secretion beginning at puberty and lasting, to some degree, for the rest of a man's life.

masculine life (Fig. 9). The male will surely need testosterone later, but without this dose so early in his history, it wouldn't be enough.

We know this largely from experimental data on rodents, so we now need to consider how far this information applies to man (and other primates). Chapter 1 describes how some species, such as rats, adopt a reproductive strategy that involves high cost to the newborn pups—most of them die—but compensates for this by producing large litters. Large litters have another consequence: the young are born very immature. Newly born rats (and cats) are blind: their visual system is not yet mature enough to function properly. Their brains are also very immature. This means that events that might happen during the much longer pregnancy of humans happen after birth in rats. One of those events is the action of testosterone on the developing brain that, in man, happens well before birth, but in rats (and mice) just after it. This allows experimental manipulation of the early hormonal environment: we can change what happens naturally and see what effect this has. It turns out to be dramatic.

Remove the testes from a newborn male rat pup, and he will never function properly as a male when he grows up, even if you were to give him testosterone. There are several reasons for this. His penis doesn't grow at puberty: it has not been sensitized to pubertal testosterone by the conditioning effect of testosterone in early life. But there are ways of counteracting this: to enable the little male rat's penis to grow despite the early lack of hormone. Does this put everything right? Not at all. The male shows rather little interest in females, despite being given adult doses of testosterone. Even more strikingly, if he is given oestrogen and progesterone, in doses that result in a normal female becoming sexually active ('oestrous') then he will show definite signs of female-like behaviour in the presence of another, normal, male. Something has happened in his brain, and it's permanent: no matter for how long he gets doses of testosterone that, in normal males, would be sufficient: he never functions as a male himself. The absence of testosterone early in life has thus had several striking consequences: it has altered the subsequent growth of his genitals; it has altered his motivation (he responds more to males than females); and his ability to display the rather stereotyped sexual postures of the normal female has been enhanced. He appears to be 'female-like'.

The opposite experiment gives confirmatory results. Give a little newborn female rat pup testosterone just after birth, and she grows up behaving rather like a male. She is resistant to the behavioural effects of oestrogen and progesterone—she doesn't become oestrous (i.e. showing the typical posture of a sexually active female in the presence of a male)—and, given testosterone, will behave much more like a male than would a normal female (i.e. she is more likely to try to mate in male-style with an oestrous female). Other signs of 'maleness' are also present: she won't show the normal 4–5 day oestrous cycle typical of the female rat, but a constant state of oestrogen secretion that recalls the similarly constant state of testosterone secretion in the male. She has become something of a male. Similar results can be shown in other species; they are not peculiar to rats. The whole field actually started with some experiments by Alfred Jost on rabbits in the

1940s: he removed ovaries from female embryos (a formidable technical achievement), and showed that they continued to develop as females, whereas if he castrated male embryos, they often became female-like. But testosterone is not solely responsible for 'maleness'. The foetal testes produces other substances, including one called 'anti-Mullerian hormone', which causes the developing uterus and its associated structures to regress. If it doesn't function properly, the little male will develop both male and female internal reproductive organs side-by-side.

These dramatic results on behaviour, based initially on the work of three Americans, Calvin Stone, William Young, and Frank Beach in the 1930s to 1950s, have obvious and huge implications for understanding human sexuality provided, of course, that they can be extrapolated to man. But at the same time that experimental data were producing such a wealth of evidence implicating testosterone in the development of sexuality, those working on humans were having ideas of quite a different kind. Human sexuality was divided into three components: gender (whether one thinks of oneself as male or female); preference (whether one finds males or females sexually attractive); and role (the behavioural pattern exhibited during sexual encounters but also in society in general). The last two can be studied experimentally in animals: the first is much harder (though there have been attempts). But the real separation between what the experimental scientists were doing, and what the human psychologists were thinking was this: human sexuality, it was suggested, was undifferentiated at birth.[†] Babies were born sexually 'neutral': so how did they become males or females? By the way their parents treated them. The external appearance of a baby told the parents that this was a 'boy' or a 'girl'.

[†] John Money (1921–2006), a psychologist at Johns Hopkins University, was highly influential in promoting this definition of human sexuality, particularly as it applies to transgender and intersexuals (hermaphrodites). He was also an advocate of sexual neutrality at birth.

They then, maybe unknowingly, treated them as such, and it was this that formulated their sexuality. This psychosocial theory had serious and obvious implications. First, it put the onus of a child's development and subsequent functioning as a sexual being on the parents, even though they didn't know it. Second, it predicted that it might be possible to change sexuality by altering a parent's attitude. As it happened, a fateful surgical error enabled this last prediction to be tested. In 1966 a circumcision went wrong, and a baby boy (one of twins) lost his penis. The psychology of the time dictated that this could be compensated by reassigning the baby as a female—removing the testes, dressing the child as a girl, and giving 'it' a girl's name. This was done. It failed. The baby did not grow up to believe 'she' was a girl, and reverted to a man later in life. There was a tragic outcome—he committed suicide. Interestingly, his twin, who had retained his penis and grew up apparently normally, also committed suicide, so maybe there were significant factors in the family's life other than the tragic story of a lost penis.

It would be easy to criticize the handling of this case, and much criticism has, in fact, been made.[18] But the prominent psychologist of the time, John Money, had well-developed arguments for his position, though history has not validated it. It is not a crime to be wrong, or most scientists would be in prison. What is wrong is not to change your mind in the face of overwhelming evidence against your point of view. But that's not too uncommon either. We might conclude that the idea that human sexuality is totally unformed at birth is not true. But the debate has not ended. A second case has been described, in which a male lost his penis (for a similar reason) at around 2 months. In this instance, 'she' was also raised as a female and accepted herself as such, though she was bisexual. It is obvious that basing firm conclusions on single cases, particularly when they suggest different interpretations, is unwise, though this has not prevented fierce positions being taken.[19]

Cloacal extrophy is another extremely rare condition in which the external genitalia do not develop. Since it is much easier for surgeons to construct female genitalia, some babies with XY and cloacal extrophy have been 'declared' to be females. Of 33 males with this condition reared as females, 42% declared themselves to be 'female' at ages 4–13 years, but 55% said they were really male. All those assigned as 'males' remained 'male'.[20] The problem with these data, besides their mixed nature, is that we do not know whether their cloacal extrophy was caused, at least in some cases, by abnormal testosterone secretion or altered sensitivity to it (e.g. variations in the androgen receptor). 'Experiments of nature' have thus been intriguing, but not decisive.

Even if it were technically possible to castrate a male foetus at around the 10th week of pregnancy, or give a female one an infusion of testosterone at around the same time, it is obvious that this would be outrageous and not ethically acceptable. So we have to rely on accidents of nature, or medical treatments with unintended consequences. The first 'accident' is a genetic disorder called 'congenital adrenal hyperplasia' (CAH). As its name implies, this is a genetic ('congenital') disorder that results in the adrenal glands, which lie over the kidney, becoming larger than normal. The adrenals normally produce several hormones, one of which is cortisol. Cortisol is essential for life. Babies with CAH lack an enzyme that normally makes cortisol from a precursor steroid. But lack of the adrenal enzyme has another consequence: instead of making normal amounts of cortisol, the infant adrenal makes testosterone (and other androgens). The brain senses this lack of cortisol and, through the pituitary, tries to make more cortisol. The adrenal gets bigger under this extra stimulation but all the pituitary does is to increase the amount of testosterone from the adrenal. If the affected baby is a girl, she is born with masculinized genitalia (and sometimes thought to be a boy, initially at least). During pregnancy her mother's cortisol has kept her alive.

She needs urgent treatment: she is given cortisol, which not only provides what she lacks, but also turns off the extra stimulation from the pituitary and hence the surge of testosterone. All very satisfactory, but she has been exposed to testosterone during foetal life.

There are early descriptions of heavily masculinized individuals who behaved like males but turned out to have a uterus, though it's not certain that they had CAH. Nevertheless, despite rapid postnatal treatment, CAH girls do show some evidence of male-like traits.[21] They play more like boys (i.e. with trucks and guns) than girls. They are more aggressive than unaffected girls. When they grow up, they show less satisfaction than their peers about being female (and a few have been reported to consider themselves male). There is an increased likelihood of their becoming bisexual. Their menstrual cycles may be less regular than usual. But this is not seen in all cases: CAH can be the result of deficiencies in different enzymes, and in some cases those deficiencies may be only partial. This implies differing degrees of exposure to testosterone. But it does look as if abnormal exposure to testosterone during foetal life—levels are increased in amniotic fluid[‡] in CAH—does affect a range of gender-related attributes in the way the experimental data might predict. There is also a second prediction from the experimental work on animals: that gender-related behaviour in boys with CAH would be indistinguishable from normal, and this does seem to be true.

A converse 'accident' of nature, even more dramatic, has already been mentioned. The 'androgen-insensitivity syndrome' (AIS) is the result of a mutation in the androgen receptor, rendering the person unresponsive to his own testosterone.[22] Babies with this condition, if it is complete, are often thought to be girls, and brought up as such. Their condition may only be discovered at puberty, when they

---

[‡] The fluid that surrounds the developing foetus.

develop only sparse pubic and axillary hair and fail to menstruate. But the important point is that, despite their possession of only one X chromosome and the presence of a Y, these girls (for that is what they are) are psychologically indistinguishable from normal women across the gamut of gender-related attitudes and behaviour. This is the strongest evidence we have that testosterone is responsible for much of what we call 'masculinity' in humans, though those favouring a 'social' role for gender determination will point out that these individuals have been regarded as girls by their parents. Why is it that there is a complete reversal of gender in AIS, but only signs of partial masculinization in female CAH? Surely, if testosterone is so powerful, then one should be the mirror-image of the other? The answer, still uncertain, may reflect the amount or timing of excess testosterone in females with CAH, which may not mimic exactly that occurring in boys (recall also that CAH is often incomplete). It seems extraordinary that a single, simple, molecule could have such a powerfully deterministic action, and no doubt there are those who would dispute that it has. But taken together the experimental evidence (which includes giving testosterone to pregnant monkeys, which also masculinizes their female offspring) and clinical research, it is hard to conclude anything else.

In Chapter 2, we described how testosterone can be converted in the tissues to a more powerful substance (dihydrotestosterone: DHT). DHT binds to the androgen receptor even more avidly that testosterone, but seems to be particularly important for the male's genitalia, including the penis and prostate (hence the use, already mentioned, of drugs that block its action as part of the treatment of cancer of the prostate). The enzyme responsible for this conversion is called 5-alpha reductase (5áR). There are rare cases in which the gene that controls the formation of this enzyme has mutated so that the enzyme is no longer effective. Little boys with this condition may be born with undeveloped genitalia, and assumed to be girls. At puberty, the increasing amount of testosterone is sufficient for their penis to start to grow, though they may need additional surgery for acceptable function. This phenomenon had been recognized by the Dominican Republic society in which a cluster of such cases occurred as the 'penis at twelve' children. The essential point is that, although some of these babies are raised as girls, at puberty they usually choose to make the transition to being a male.[23] So the conversion of testosterone to DHT does not seem important for the way the brain develops its 'male' characteristics. There is separation, it seems, between the prenatal hormonal control of the brain and the genitalia.§ 5áR deficiency also argues against gender of upbringing being a major determinant of later sexual identity.

There are aspects of human sexuality that are not easily studied experimentally: we have mentioned gender identity (whether you think of yourself as male or female) as one. Homosexuality is another. CAH females may have an increased incidence of bisexuality (though not everyone who has studied this agrees). What about male homo-sexuality? There is a logical argument here: if male attributes, including being sexually attracted to females, are testosterone-dependent, then might it be the case that testosterone deficiency of some sort at some stage in prenatal life could be responsible for males being attracted to other males? The initial research, which looks naïve today, was to measure testosterone levels in adult homo- and hetero-sexual men (the idea that they could be separated in this binary way was also naïve**). The proposition was that homosexuals were less 'masculine' than heterosexuals, and this would be reflected in lower

§ *Middlesex*, by Jeffrey Eugenides (HarperCollins), is a novel in which a Greek-American man with 5-á-reductase deficiency is the central character. Inventive, original and illuminating, it deservedly won a Pulitzer Prize.

** It was Alfred Kinsey and colleagues who introduced the idea that human sexuality lies along a scale from 0 (exclusively heterosexual) to 6 (exclusively homosexual) with intervening grades of bisexuality. 'Males do not represent two discrete populations, heterosexual and homosexual. The world is not to be divided into sheep and goats. It is a fundamental of taxonomy that nature rarely deals with discrete categories...The living world is a continuum in each and every one of its aspects.' (1948), *Sexual Behavior in the Human Male*. Saunders, New York.

testosterone. Would you believe, the first studies found exactly that, to great acclaim. Then someone (it was the distinguished behaviourist Frank Beach,[††] no less) pointed out that the heterosexuals had been recruited from middle-class professionals, whereas the homosexuals had been recruited from gay clubs, in which smoking cannabis was then rife (this was in the 1980s). Cannabis lowers testosterone—an excellent example of not controlling one's variables! There is no such evidence in more careful studies, though we should mention here that testosterone levels can be altered by all sorts of behaviour and circumstance. We will return to this again.

There remains the question of prenatal testosterone: could this play a role? Could levels below a certain critical value increase the possibility that a male baby might become gay? It was not until 1967 that homosexuality was decriminalized in the UK, and not until 1973 that is was declassified as a mental disorder by the American Psychiatric Association. But in 1785 Jeremy Bentham had written:

> To what class of offences shall we refer these irregularities of the venereal appetite which are styled unnatural? When hidden from the public eye there could be no colour for placing them anywhere else: could they find a place any where it would be here. I have been tormenting myself for years to find if possible a sufficient ground for treating them with the severity with which they are treated at this time of day by all European nations: but upon the principle utility I can find none.[24]

At that time homosexuality attracted the death penalty.

If a society condemns homosexuality, then one consequence is that society will try to prevent it, either by sanction or, in the case of East

---

[††] Frank Beach (1911–1988) was a towering figure in the experimental study of sex behaviour in rats. He made fundamental contributions to our understanding of the way that hormones controlled sex behaviour, and influenced development. Characteristically, he wore a string bow tie and looked like a Mississippi gambler (he was actually from Kansas). His obituary says 'he knew how to party'. He wrote a famous book, *Hormones and Behavior*, in 1948.

Germany before reunification, by science. Eliminating homosexuality was seen as a prime objective. Glory awaited any scientist that achieved it. It was hypothesized (without any real evidence) that homosexuality in men was caused by low testosterone during embryonic life. It was therefore proposed that all pregnant women carrying boys should be tested for testosterone; those falling below a certain level should be aborted or given extra testosterone. This policy received widespread criticism from the rest of the scientific community outside communist Germany (the proponents remained unmoved) and was not actually adopted. It is another example of the fearful results that can occur if science is either misunderstood or misinterpreted. It is also a good example of how science has always to be on guard against social and political pressures that are incompatible with its own standards: this is particularly relevant in the high-octane field of human sexuality, where science merges not only with politics and social mores, but religion, class, folklore, tradition, and even prejudice.[25]

Where does this mixed evidence leave us? What can we conclude about the determinants of human sexuality and, in particular, the role of testosterone? It is clear that human sexuality is complex, and not definable as a single entity. If we limit ourselves to the three dimensions of sexuality set out above, clinical evidence shows that each can vary independently. For example, a transgender (identity) can be in either direction, can prefer either sex, and take either gender role: all combinations can and do occur.[26] This implies that each may be determined separately. The distinction is not always obvious. For example, a unique long-term study of small boys who cross-dressed and behaved like girls revealed that they didn't become transgenders, as might have been predicted, but gay.[27] There is no doubt from the experimental and clinical evidence that abnormalities in testosterone levels during intrauterine life could affect any or all of these aspects of human sexuality. Different levels or, more likely, different timings of prenatal testosterone could alter each separately, though there is no clear evidence on that point. Since AIS individuals behave like heterosexual girls, this does suggest that testosterone may have a lot to do with the development of masculine heterosexuality.[28] This does not mean, however, that lack of testosterone is necessarily implicated in the development of homosexuality. Other factors (e.g. oestrogen levels) might interact with testosterone and result in distinct patterns of sexuality. And then there are genes: if one of a pair of identical twins is gay, then the chances that the other is also gay is around 50%. This means that while genes may have a powerful influence on the likelihood of a male becoming gay, this is only part of the explanation, the rest being some other element within the individual's environment. There have been claims that an altered gene (a 'gay gene') has been identified: but these have not been substantiated. If genes might explain around half of an individual's sexuality, what in the environment might explain the other half? Are there definable elements in the complex social and physical environment experienced by every child that might be an influence? Freudian psychologists have long talked about domineering mothers and/or absent fathers. There is no firm scientific evidence so far to support these ideas. As far as testosterone is concerned, while we lack details, the principle is clear: prenatal testosterone has a plausible role as a powerful influence (probably only one) on the development of human sexuality.

But we cannot rule out other, non-hormonal factors. Though it seems unlikely that the role attributed to parents in the development of sexual identity is defensible, there is a pile of evidence showing that early adversity—for example, parental neglect, hostility, or actual abuse—can have profound and long-lasting results for the child's later mental state. This is likely to include aspects of sexuality as well as other psychological traits. Most attention by workers in this field has been given to the way that adversity early in life results in later mental illness (e.g. depression or conduct disorder) rather than sexuality. Parental maltreatment may of course be a result of homo-sexuality in children rather than its cause. Whether parental behaviour can itself predispose to particular sexual attitudes, or combine with other factors, is still not really understood. It is far too common to take up an exclusive position on the side of either 'biology' (e.g. testosterone) factors or 'psychology' (including early experiences) in the development of human sexuality. This is more a consequence of the so-called expert's training, background and prejudice than a rational approach to the fascinating question of why we are what we are in our sexual lives. The two sets of ideas are simply different angles on the same viewpoint, and they interact powerfully.

To understand the role of prenatal testosterone in humans—not only on sexuality, but on all the other trappings of masculinity—one needs an accurate measure of individual differences in exposure during embryonic life. This is clinically and ethically impossible. There currently is no way of directly monitoring testosterone levels in the foetus reliably and safely. But a clue was available for all to see: one of those discoveries that anyone could have made very simply and without any technical sophistication. I have no doubt that many people noticed it, but it wasn't reported in the scientific literature until 1875. If you are a male, turn the palm of either hand towards you, and look at the length of your fingers. You are likely to see that your fourth finger is a little longer than your second. If you are a woman, then either the two fingers are about the same length, or the second is a little longer than the fourth (Fig 10B). Now, this varies, and it's not true for all males or females. But if you measure the relative lengths among, say, 100 men and women, then you will get a statis-tically significant difference in the ratio between the two fingers (it's called the 2D:4D ratio: D = digit) between men and women. Were you to plot each individual, you would see that, although this difference is quite obvious overall, there is a very considerable area of overlap: that is, some men have a higher ratio than some women, though in most

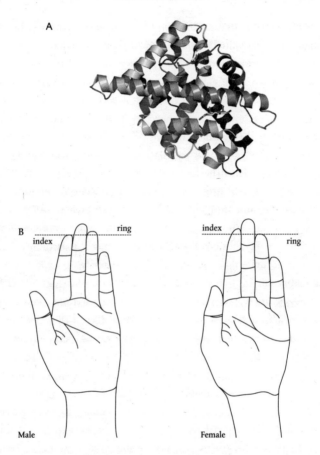

Fig. 10. (A) The complex protein that is the androgen receptor. It is made up of a long ribbon of amino acids. Testosterone attaches itself to this molecule. (B) The different ratios between the lengths of the second and fourth digits in males and females. Note that these ratios, though statistically different, overlap.

men it's less than in most women. You can't confidently predict the gender of an individual by measuring his/her digit ratio.

It wasn't until 1998 that John Manning suggested that the 2D:4D ratio might represent early exposure to androgens. Subsequent evidence has supported this, but only to a degree. The sex difference in digit ratio is present before birth, so it's dependent on events during intrauterine life. You will recall that AIS individuals, though possessing a Y chromosome, are unresponsive to their own testosterone because of a genetic defect in their androgen receptor (Fig 10A). Their digit ratios resemble those in females. CAH girls have been exposed to excess testosterone during embryonic life: they have male-like ratios. Now for some intriguing, and controversial data: lesbian women overall have more male-like ratios than heterosexual ones. But in males there is no difference between gay and straight men. Taken at face value, this suggests that lesbianism is more likely if a female foetus is exposed to increased levels of prenatal androgen (this agrees with the increased incidence in some reports of CAH women), but that variations in prenatal testosterone play no discernible part in the development of homosexuality in men. But there are problems.

The first is that there is no biological reason why these two fingers should respond to prenatal testosterone. It is surely of no importance for their function. The second is that, like any measure, we have to be concerned about the sensitivity of our measure (how good is it in detecting difference?), how specific is it (does it measure only what we think it does?), and how accurate (is the value we get a true one?). There are questions about the reliability of the digit ratio as a measure of early exposure to testosterone on all three counts. Factors other than testosterone may influence the digit ratio (though none have yet been found, so far as I know). While the strength of the ratio is that it may reflect total exposure to testosterone (not just the amount of hormone, but sensitivity of the androgen receptor), it may not be able to do this with the accuracy that some have assumed. You can't, for example, tell whether a woman is gay or not by measuring her digit ratio because the difference between gay and straight women is too small and overlaps too much. The important conclusion is that while the digit ratio may be a useful guide to differences in early exposure to testosterone between different groups (e.g. gay vs. heterosexual women, CAH females vs. normal ones), it may be much less useful for assessing individual difference within groups and assuming these reflect corresponding differences in embryonic exposure to testosterone. But many studies try to do exactly that. Attempts to relate the digit ratio to variations in the androgen receptor (and hence sensitivity to testosterone) have failed.[29] The digit ratio remains an intriguing snippet in the testosterone story: the real problem is—it's so easy to measure that anyone can do a study, and assume they are assessing accurately their subjects' early life testosterone exposure.[30] All you need is a photocopier and a ruler.

Early exposure to testosterone may influence aspects of a male's behaviour other than his sexuality. As we have seen, there are marked gender differences in play behaviour in young children. There are also marked gender differences in the interests of adults, although, as with all such differences, there is an overlap between the sexes. Males tend to be interested in objects (e.g. cars), processes (e.g. computer programming), or events in the physical world. Most of those who choose information technology as a career are male. Women tend to be more interested in occupations that involve communication with others. Thus most neurosurgeons are male, but many psychiatrists are female. If we can discount systemic barriers to either gender in these professions, they seem to represent gender-biased choice. That is not to deny that such barriers may still exist: this is certainly true, more in some cultures or occupations than others. But even in a utopian state in which there are none, there are likely to be distinct sex differences in preference or abilities for particular occupations or pursuits. Those that demand equal numbers of either sex in a given occupation sometimes confuse equality with similarity. Many of these gender-differentiated psychological attributes can easily be related to corresponding differences in reproductive strategy—males need to take an interest in the physical world for hunting, competing, etc.; women have emotional requirements during child-rearing, and so on. It is likely that testosterone plays a role in promoting male-type psychological tendencies. The evidence from CAH females would suggest this (see earlier in the chapter). But note these are only tendencies; there is much overlap between the sexes, and much individual variation within each sex.

But suppose we were to imagine a person who had an overdose of 'male-like' characteristics. Such an individual would be intensely interested in objects, such as machines or structures, or processes such as drawing or using computers but have little ability or wish to communicate or interact socially with others. There are such people, and they are classified as being autistic or, more precisely, having an autistic spectrum disorder (ASD).[31]

ASD males outnumber females by about two to three times (but note that autistic females do occur). The fact that they appear to show exaggerated male-like behaviour—including heightened aggression—has prompted the suggestion that some ASD may be the consequence of over-exposure to testosterone during early intrauterine life.[‡‡] Attractive as it may seem, there are problems with this idea. The first is lack of empirical evidence: attempts to show increased levels of amniotic (foetal) testosterone in babies subsequently developing ASD are still not decisive. Furthermore, CAH boys are not more likely to be autistic than other children. CAH girls are said to have more autistic-

[‡‡] Recently, there is evidence that not only elevated testosterone but also a variety of other steroids during gestation may be associated with later ASD. Possible mechanisms for this (rather unexpected) and complex result and whether it contributes to ASD are still to be unravelled. S Baron-Cohen et al. (2014), 'Elevated fetal steroidogenic activity in autism'. *Molecular Psychiatry*, vol. 20 pp. 369–376.

like traits,[32] but these are actually another reflection of their more male-like psychology (see earlier). Whether differences in sensitivity to testosterone (e.g. variations in the androgen receptor, or in genes known to influence response to steroids) might be implicated has not been reported.[33] If one identical twin has ASD there is about an 85% chance of the other being affected, which points to a genetic mechanism. But there are other problems with the testosterone idea. About half those with ASD have mental handicaps, and testosterone would not be expected to be responsible for these. Others may have either normal or outstanding (but limited) mental abilities—often classified as Asperger's syndrome—but lack emotional responses or insight into the mental states of either themselves or others (i.e. a 'theory of mind'). Other physical abnormalities are also common in ASD, and these are also difficult to attribute to excess testosterone. If excess embryonic testosterone were really involved, then one might expect altered frequencies of sexually related traits, such as homosexuality or transsexuality: the evidence for this is inconclusive. The wide range of symptoms in ASD make it difficult to pin down a corresponding brain abnormality, though a disturbance in neural connectivity of some sort is suspected. Perhaps excess testosterone could be implicated in one of the subtypes of ASD, but this is entirely speculative.

But let's not allow all these scientific 'ifs' and 'buts' to blind us: the evidence that testosterone operates on the body and the brain of a male from his earliest existence is overwhelming. While the slings and arrows of later life may have all sorts of extra effects, and this includes testosterone itself during adolescence and adulthood, the pattern has been laid down: the door to a testosterone-driven life is open, whether it be heterosexual, bisexual, or homosexual. The trajectory of a man's life can take many forms: some dependent on his personal qualities and experience, some on happenstance, but others on influences quite outside his control or even knowledge, such as geopolitical events that shape a generation.[34] Our little XY embryo may be a male because of his Y chromosome but only because this chromosome results in testosterone being secreted during a critical period of his early life. Without this testosterone, there is no male. Now he waits for the next great testosterone-driven event: sex.

4

Testosterone and Sex

No single physiological variable ... is as important in determining the occurrence or level of sexual responsiveness as the amount of gonadal hormones in the blood. This kind of relationship is unusual if not unique in behavioral physiology. In no other area of behavior do hormones appear to occupy such a commanding role, and no hormones have nearly as important an influence on sex behavior as the gonadal hormones.

G. Bermant and J. M. Davidson (1974), *Biological Bases of Sexual Behavior*. Harper and Row, New York

Sex is an antisocial force in evolution. Bonds are formed between individuals in spite of sex and not because of it. Perfect societies ... societies that lack conflict ... are most likely to evolve where all of the members are genetically identical. When sexual reproduction is introduced, members of the group become genetically dissimilar ... The inevitable result is a conflict of interest. The male will profit more if he can inseminate additional females ... Conversely, the female will profit if she can retain the full-time aid of the male ... The offspring may increase their personal genetic fitness by continuing to demand the services of the parents when raising a second brood would be more profitable for the parents.... The outcomes of these conflicts of interest are tension and strict limits on the extent of altruism and division of labor.   E. O. Wilson (1975), *Sociobiology. The New Synthesis*. The Belknap Press, Cambridge, MA

A great wedding breakfast was prepared. Cupid reclined in the place of honour with Psyche's head resting on his breast.... Jupiter was served with nectar and ambrosia by apple-cheeked Ganymede, his personal cup-bearer; Bacchus attended to everyone else. Vulcan was the chef; the Hours decorated the palace with red roses and other bridal flowers; the Graces sprinkled balsam water; the Muses chanted the marriage-hymn to the accompaniment of flute and pipe-music...Venus came forward and performed a lively step-dance in time to it. Psyche was properly married to Cupid and in due time she bore him a child, a daughter whose name was Pleasure.

Apuleius, *The Golden Ass*. Translated by Robert Graves.
Penguin Books, Melbourne

S ex is entirely in thrall to hormones. Let's begin with females of species other than man or the other primates. Remove the ovaries of any female rodent, feline, marsupial, and many other mammalian species, and she will never mate again. Not unless she gets given a shot or two of the right hormones. There are good biological reasons why the sexuality of the females of most mammals is so tightly controlled by hormones. As we have already seen, reproduction is a highly demanding, risky, uncertain business for females. Carrying young makes many demands: on her nutrition and the necessity to build a nest, for example. Giving birth is not without its own danger. Then come further, huge, metabolic demands from nursing the young, and the risks of defending them against rivals or predators. She has to find extra, nutritious food to keep both herself and her newborn alive. Whether her species goes the way of the rat, and gives birth to a crowd of immature babies, or the way of the deer or the sheep and produces one or two much more mature infants, it's a biological gamble. She can maximize her chances in several ways: one is to ensure that birth occurs at the right time of year, usually the spring (in temperate parts of the world) when new and more plentiful food is becoming available. Another is to ensure that she has mated with a male that has the qualities that promote the survival of her young.

Hormones are highly suitable for enabling her body to exercise strict control over reproduction. By making her brain totally dependent on them she ensures that when her ovaries contain ripe eggs, they also secrete the right hormones and so she becomes sexually active at the best time. So her ovaries, the source of fertility, are also the regulator of her sexuality. But here is a problem. Ovaries are buried deep inside her abdomen. They have no access to the environment; they cannot 'know' when it is spring. What tells them is a signal from the pituitary, a gland lying just underneath the brain. It, too, sends out a hormonal signal, not a steroid, that wakes up the dormant ovary, and causes both its eggs to ripen and steroid hormones such as oestrogen to surge into the bloodstream. But this only deflects the problem. The pituitary, deep in the skull, can't access the environment either. So what tells the pituitary that spring has arrived?[35]

Just above the pituitary lies a tiny part of the brain, the hypothalamus. Size is no indicator of importance, and this part of the brain is very important indeed. It has two major functions: it detects what goes on in the body—the levels of glucose, or water, or sex hormones, for example. It also sends its own signals (more hormones) that tell the pituitary what to do in case of need. The hypothalamus is buried deep in the brain, just behind your eyes. The most reliable predictor of imminent spring is day length: if the days are getting longer, you can be sure spring is on its way. The hypothalamus does have access to light, since nerve fibres from the eyes connect to it (this regulates daily rhythms, not reproduction). Light signals also go to another gland, the pineal, also buried deep in the brain, but able to calculate day length. It too joins in the hormonal cascade, sending its own signal (melatonin) to the hypothalamus. A shorter melatonin signal means a short night, hence a lengthening day. The hypothalamus tells the pituitary; which tells the ovaries; which initiates fertility and, by secreting steroids, sexuality. In species with a short pregnancy, a shortening melatonin pulse is the signal. In those species that mate in the autumn (sheep, for example: they carry their young until the spring) the hypothalamus responds to a lengthening melatonin signal, representing the end of summer. This chain of command, despite the large species differences in the exact way that the females' reproduction is carried out, regulates breeding tightly and highly effectively. So it is that females of nearly every mammalian species, if they live in temperate climates, breed in the spring. The final, essential, part of this control system is totally dependent on the steroid hormones from the ovary.

Even during a breeding season, the females of most species remain in thrall to their hormones, becoming sexually active only periodically, at times dictated by cyclic changes in the secretion of hormones from their ovaries. A female rat is only 'in heat' for a few hours every 4 or 5 days: her pituitary regulates the cyclic release of hormones from her ovaries, and these act back on her hypothalamus to initiate her sexual activity. Sexuality in ewes is likewise periodic, and this is also dependent on intermittent secretion of ovarian hormones.

We have already seen the reasons why a book on testosterone bothers so much with how female reproduction is controlled. A male's 'fitness' can be quite simply defined: by the number of offspring he produces that themselves become fertile and produce yet more young. So an ideal male, you might think, is one that can sire most young.[36] But here's the catch: as we have seen, the risks of reproduction itself are taken, in most species, by the female (but not the process of getting a mate: that represents risks for the male). It is her ability to conceive, carry a pregnancy, deliver and nurture her young, that will constitute the male's fertility, and hence his fitness. He depends on her, so he has to adapt to her needs. His sexual behaviour has to maximize the chances of a fertile mating, and this will vary according to the females' reproductive physiology. His penis has to deliver semen in the optimum way, which may depend on her anatomy. The pattern of mating varies markedly in the males of different species. Some (e.g. dogs) have prolonged single intromissions; others (rats, some monkeys) intromit quite briefly a number of times before ejaculating. Men usually have a single intromission with a series of thrusts, but, unlike other species, can vary their patterns of copulation according to circumstances. The biological reasons for these species differences remain unclear in many cases: in some they seem to be related to the need to provide sufficient stimuli for the female to ovulate.

And he has to be fertile and sexually active when she is; there's no point at any other time. Reproduction is a risky time for the male as well, though the risks facing a male are very different from the hazards that a female has to overcome. But we cannot fully understand the complexities of male sexuality without taking that of the female into account.

No surprise: testosterone is the essential regulator of both fertility and sexuality in non-primate males. Somewhat before the female begins her breeding season, the male begins his; he has a lot of preliminary issues to sort out. His sexual activity begins for much the same reasons: altered day length. The chain of command is activated. Testosterone surges out of his awakening testes, and sex takes over his life. But there are some odd differences. Females of these mammalian non-primate species are so sensitive to their hormones that, if you remove their ovaries when they are in 'heat' ('oestrus') within 24 hours or so they become entirely and permanently asexual. Not so the males. Remove his testes, and there is a slow decline in his sexuality, but it will take a long time—maybe several weeks or even months—for his sexual activity to disappear entirely (and it may not). We could, of course, suspect that this is because testosterone remains in his blood: not so. Measure it, and it disappears as quickly as oestrogen does in the female. For a time, it seems that his brain can continue to remain 'sexual' without testosterone, but not forever—something we need to consider further and, if possible, explain. Interestingly, a similar gender difference can be seen if one gives hormones to castrated males or females. The latter tend to respond almost immediately, sexual activity appearing within a day or two of first treatment (or even a few hours). But the response in castrated males is much slower, and full and final levels of sexuality occur only after days or even weeks of testosterone treatment. There's something intriguingly different about the way that hormones control sexuality in the brains of the two sexes, and we'll need to discuss how they act on the brain if we are to understand it (Chapter 10).

How far do these principles apply to primates? Before we look at humans, we should consider what information we can glean from our closest biological relatives: prosimians, monkeys, and apes. The prosimians are species that, because of their relatively less complex brains and distinctive appearance, are considered 'primitive' primates, in the sense that they have features that are closer to what we assume might be earlier, prehistoric primates. The lemurs of Madagascar and the lorises of Asia are prosimians. Monkeys are divided into New and Old World species, and these inhabit South America or Asia and Africa respectively. Apes are divided into 'lesser' (gibbons and siamangs) or 'great' (orang-utans, chimpanzees, and gorillas). Many primates live at or near the equator, which itself makes a light-driven annual breeding season unlikely. So evidence of seasonal changes in testicular activity, and hence testosterone secretion, is lacking or incomplete in many species. Despite this, many prosimians do have annual bursts of reproductive activity. Monkeys are a mixed bunch. In some, for example the rhesus monkey of India, there is an annual season, though in the autumn rather than the spring (since they have long pregnancies); other species breed throughout the year. Some seem to respond to rainfall rather than light, so timing the births of their young to coincide with a more verdant landscape in the rainy season. Since monkeys are generally vegetarian, the state of fruit and foliage are powerful influences on reproduction. Apes are generally non-seasonal, though the caveat about their habitat has to be remembered. While we can impute changes in testosterone as a regulator of such seasonal changes as do occur by observing alterations in the size of testes, the only really solid evidence comes from castrating male primates and then giving them testosterone. We consider human sexual behaviour in more detail later: here it should be mentioned that, since *Homo sapiens* originated from Africa, not a temperate climate, the need to regulate breeding in response to changing day length is not part of our evolutionary history, any more than that of other primates. Had early primates been able to spread further from the equator, the story might have been different.

There have been relatively few experimental studies of the effects of castrating male primates on their sexuality compared to the enormous literature on non-primates, particularly rats. Nevertheless, the results seem similar. Sexual activity declines slowly but progressively over many months or even years, and may never disappear entirely. Testosterone restores behaviour: but here is a foretaste of a more complex situation, to be explored more fully later. Although the basic physiology of different species of primates seems quite similar, the arrangement of the social groups in which they live varies widely. This has powerful effects on sexuality. In some primate societies, the males form a hierarchy based on aggression, fights, and defeats (Chapters 5 and 6). If all the males in such a group are castrated, then giving testosterone to the top male restores his sexual behaviour. But in the lower males: nothing. Something other than a simple hormonal control of sexual behaviour is operating. The social control of sexuality is all-powerful (and not only in primates). Variations in reproductive strategy in different non-human primates owe much more to the different kinds of social groups in which they live than to any hormone. There's a lesson there for humans, as we'll see.

So now for man: there are those who think information from animals has little relevance for understanding human sexuality:

Our standards of human sexuality are especially warped, species-ist, and human-centric because human sexuality is so abnormal by the standards of the world's thirty million other species.

Jared Diamond (1997), *Why Is Sex Fun?*
Weidenfeld & Nicolson, London

So how far do we carry into our world, so different from that even of our nearest primate relatives, the hormonal control over sexuality so clearly apparent in them and other mammals? Has social evolution carried with it a corresponding biological change, or does so-called modern man still carry with him the physiology of his primeval ancestors? It's a critical question.[37] If the evolution of the great human brain has exempted it from hormonal control, then we will need to consider how this could have happened. One implication is that the control of human sexuality will have passed on to other agents. On the other hand, if, despite the complexities of the human brain, the basic, primitive hormonal regulation of sexuality remains in the background, how does this fit in with the development of complex social and other rules that regulate sexuality in man? There are those who suggest that the development of the human brain, enabling more complex behaviour and technical achievement than any other species, has also liberated us from the control of hormones to which they are shackled. Is this an established fact, or simply human hubris? Humans don't have breeding seasons, though there may be (rather small) peaks in births in temperate climates in autumn and a lesser one in spring.

Women generally do not show the marked sexual variation in phase with their menstrual cycles that characterizes some other species (including some monkeys), though there are plenty of studies demonstrating that some women, at least, do report sexuality peaking near the middle of their menstrual cycle diminishing during the latter part; but others deny this. Yet others claim a peak just before menstruation (see Chapter 9 for further discussion). Cyclic variation in sexuality in women is certainly not as marked or universal as in, say, female rhesus monkeys, let alone rats. Social inhibitions, such as a taboo on sex during menstruation, may accentuate or even be responsible for some of the evident periodic change in women's sexual activity, rather than being a direct effect of hormones. The menopause represents a natural removal of oestrogen and progesterone, and reports of diminished sexuality are common. Does this imply hormonal control? There are caveats. John Bancroft, a noted authority on human sexuality, has written:

> In a species such as the human, where sexual activity is far in excess of that required for optimum fertility, the woman's degree of behavioural responsiveness to her reproductive hormones will have little bearing on whether she reproduces herself optimally though her life span. The human male, on the other hand, who tends to initiate sexual pairings, has both his sexual appetite and his fertility dependent on the same hormone. The result...is greater genetic variability of hormone-behaviour responsiveness in women than in men...[38]

Withdrawal of oestrogen may lead to genital atrophy and dryness, and this may inhibit sexuality. Long partnerships, with ageing males, may have similar effects: sexual attractiveness in both sexes tends to diminish with age. The post-menopausal ovary can, in some cases at least, continue to secrete testosterone, and this hormone, surprisingly, plays a very significant role in the human female's sexuality (see Chapter 9). Testosterone can heighten women's sexual interest ('libido')—it has a similar effect in female monkeys (but not rats). It may sustain post-menopausal sexuality, at least in some women. This fact makes definition of testosterone as a 'male' hormone indefensible. It's the brain that defines what a hormone does.

Curiously, the evidence that hormones play a major role in human sexuality comes from males, rather than females. Eunuchs have been known since the dawn of human recorded history.* Men were

---

* 'To remove the testes is a great restorative, in several ways...In the Chinese court, eunuchs were known to be long-lived (the last in the Imperial Court survived to ninety-three) and their Western counterparts do just as well. In the United States, in the 1930s, orchidectomy was used, without much thought, as a treatment—or punishment—for masturbation or for minor crime. Forty years later, a follow-up of such unfortunates still resident in mental homes found that, on average, they lived thirteen years longer than their unaltered fellows.' S. Jones (2002), *Y: The Descent of Man*. Abacus, London.

castrated as punishment, as part of religious practice, and for social or political reasons. Chinese emperors from the very beginning of dynastic history vested power in those they could trust: eunuchs, because they had no family, fitted this bill. By the middle of the seventeenth century there were thousands in imperial service, and voluntary castration was common by those otherwise excluded from the highest levels by class or poverty. The Pardoner in Chaucer's *Canterbury Tales* was, it appears, a eunuch: he had thin hair, glaring eyes, a high voice, and no beard. The Romans considered men as central, perfect, and complete; women were the opposite. These cultural notions were upset by the presence of eunuchs, because of their gender ambiguity. The presence of eunuchs constantly tested the division between men and women in the later Roman Empire. But Roman medical observers made an interesting discovery: men who were castrated before puberty retained the appearance of sexual immaturity: no facial or body hair, and no deep voice. And they had no sexual activity or desire—we should recall here that a eunuch sometimes lost both his testes and his penis, although this was not common in Roman times. But castration after puberty was different: eunuchs retained their secondary sexual characters, and often some sexual activity (it is said that some members of kingly harems, guarded by eunuchs who were supposed to have no sexual interests, took advantage of this fact). This observation has been repeated many times: castration was common until relatively recently (e.g. to retain a singing voice). Just as exposure to testosterone early in life has longstanding effects on the developing brain, it seems that the resurgence of testosterone during puberty might have similar, though distinct, persistent consequences.

Before we go on to consider the role of testosterone in the adult man, there is one other curious fact. Imagine yourself looking into a pram at a 3-month-old baby boy. He looks like any baby should: appealing, attractive, and very immature. What you may not know is that his testosterone levels are, at this moment, rivalling his father's. For reasons that remain quite obscure, starting about 2 weeks after birth and lasting for about 4–6 months, a little boy's testes burst briefly into life.[39] They don't make sperm but they do secrete testosterone—in quite substantial amounts. Juvenile male monkeys do the same, but not, so far as we know, those of any other species. The burst of activity only lasts a few weeks: then the testes go back to sleep until puberty, many years later (in man). No one knows why this brief exposure to testosterone occurs, though it may have an effect on the growth of the penis. The scanty human evidence is not helpful. There is one case of a baby boy accidentally castrated at birth (so he didn't experience his postnatal surge): he grew up normally, it seems, though, of course, he needed testosterone supplements at the expected time of his puberty. An experiment was conducted on baby male marmoset monkeys who were castrated just after birth. Their behaviour during infancy, and their subsequent sexuality remained normal (if they were given testosterone).[40] The postnatal testosterone surge in man remains understudied and quite mysterious, though there is a recent suggestion that it may influence a male's preference for male group behaviour (see Chapters 5 and 8) and promote higher levels of activity but less empathetic (more autistic) characteristics.[41]

What about adults? Is the sexuality of the adult male dependent on testosterone? It is a truism to say that human sexuality is complex: even limiting this to what we might term 'primary sexuality'—that is, features directly related to a male's sexual activity. Primary sexuality can be divided into information processing (the recognition of a potential sexual partner); incentive motivation (the desire to have sex); central arousal (sexual interest in someone particular); and genital arousal (penile erection). Testosterone could have separable actions on all or any one. There have been insufficient data to decide this reliably, though there have been claims that, for example, erectile ability is more sensitive to testosterone than sexual interest.[42] Most information comes from men who have low or absent testicular function (for a variety of reasons), or are treated with 'anti-androgens' (androgen receptor blockers) for prostate cancer. Limited though it is, overall the evidence is clear: the human male's sexuality is as dependent on testosterone as that of any other primate. Males with low testosterone regain their sexuality after treatment with testosterone (but over several months). If testosterone is withdrawn, their sexuality decreases slowly over time as would be expected. Anti-androgen treatment has a similar effect: a slow decline, over many months or even years in most (but not all) men. Of course, the presence of a serious cancer will have its own consequences for sexual behaviour, which complicates interpretation. Some countries have treated male sexual offenders with drugs that either inhibit or block the action of testosterone: the results have been similar, though not reliable enough to warrant this approach (which has other, ethical, problems). Altogether it is apparent that the human male has not escaped being dependent on testosterone for effective sexual activity, though there are interesting individual differences in hormonal sensitivity which remain mysterious. Without our father's testosterone, none of us would exist.

Testosterone not only makes males seek sex, it can also make them more sexually attractive. Females play a powerful role in reproduction through sexual selection (see Chapter 6), and hence patterns of assortative mating,[43] so sexual attractiveness to women is vital. We discuss the importance of social roles, warrior-like features, and sexual display in a later chapter (Chapter 6); here we only need to note that many of the features of a male that mark him as someone who would successfully defend his patch and provide a secure home for the female and her young, are those that the female finds attractive. They are also those, like bright plumage, or wide shoulders, or prominent muscles, that are dependent on testosterone. These primeval biological features of sexual attractiveness have been modified by humans. Clothes are used as sexual enhancers: for example, padded shoulders, epaulettes, tall hats. As the complexity of assets and their value has increased, so too has the role they play in sexual attractiveness. While an earlier generation might have valued muscularity and height (which are still potent factors), social standing or wealth has become significant sexual attractants. Not all famous footballers or grand prix drivers are very pretty, but their wives or girlfriends usually are. How many times have you seen an attractive young blonde clinging to the arm of a much older, but very rich, man? Economics has invaded sexuality in humans (dowries are another example), the human equivalent of the size of a male's territory or the structure of his nest. As in so many other contexts, human beings have taken a basic biological concept and elaborated it in ways unknown to other species, but serving the same underlying function.

We need, if briefly, to consider the relation between sex and bonding or attachment. Testosterone activates sexual desire, which some would call lust, in a general way. Adolescent boys become interested in females, particularly in any attractive female. Every advertising agency knows that men are lured to objects such as cars by pictures of pretty girls. The presence of these girls says nothing about the quality of the cars though there may be a subliminal message that the car would enhance the male's attractiveness to women. But added to sexual motivation and interest, a general quality, there is a more particular one, which we call romantic attachment. This focuses the sexual attention of a man on a particular woman,[44] and, of course, is the basis of sexual selection and monogamy. It is clear that bonding occurs outside sexuality: hence the close attachment of a parent to his or her children, or to a sibling, or to a friend. Does testosterone simply enable individual sexual attachment to occur, without specifying the person to whom it will be directed? The issue is a complex one: for example, it is well-known that a relationship that begins with intense sexuality may progress to a less sexual but more emotional one, no less intense, but with possibly different psychological and therefore neurobiological mechanisms.[45] Studies on the human brain (see Chapter 10) seem to show that areas associated with reward, rather than those on which testosterone acts directly, are implicated in romantic attachment.[46] Chemical systems other than testosterone may be involved in the brain, such as the neurochemical dopamine, or the hormone oxytocin: the former part of the system that is thought to enable the brain to recognize and react to rewards (but see Chapter 10 for a caveat), the latter first shown to promote maternal bonding with the young, but now thought to be implicated in other, related properties such as monogamy, trust, and reciprocity (see Chapter 8 for a more detailed account of oxytocin). How can we fit testosterone into this schema? Maybe it is plausible to suggest that testosterone sets the scene: primes a man for sexual interest and activity (whether towards same or opposite-sexed partners), but other factors determine the direction, specificity, and nature of that activity.[47] If this is so, then testosterone might be essential but not sufficient for the sexual bonding, whether short-term or longer, that characterizes much, but certainly not all, of the human male's sexual behaviour and which poets, rather than scientists, call love.[48] But the message here is that testosterone does not act alone, and to understand its true function requires us to recognize that there are many bodily systems, myriad chemical signals, and a huge and complex brain all implicated in human behaviour, including the larger-scale consequences of testosterone. Because we focus on testosterone, we need constantly to bear this in mind.

As males age, testosterone levels in the blood decline, though not to the same extent in all men. SHBG levels often increase with age. This will reduce the amount of 'free' testosterone available to the tissues (including the brain). So measuring testosterone alone may not give a complete picture. Although there has been much talk of an 'andropause', the male equivalent of the female's menopause, the two are really not very comparable. Male fertility does not stop suddenly at around 50; in fact, many men have proved they are fertile in extreme old age. Testosterone levels do not precipitously decline over a few years, to disappear almost entirely, as ovarian hormones do in females. But many men experience reduced sexuality as age advances. There may be various causes. A male's partner is ageing as well, and this may contribute. His sexual attractiveness may decline. Erectile potency diminishes, and this is not necessarily improved by administering testosterone (though if levels fall very low, it can help). The fact is that most males' testosterone levels seem to be way above what is actually needed for optimal sexuality, though in saying this, we must always remember that response to testosterone varies among men, so levels are, as has already been emphasized, only a first approximate guide to overall testosterone activity (Chapter 2). Whether and when to treat ageing males with testosterone is being hotly debated.[49] Perhaps not surprisingly: the debate about hormone replacement treatment (HRT) in females, in whom the picture would seem a lot clearer, is similarly still unsettled. Giving testosterone has risks (as does HRT); here is a classic situation in which the risk/benefit ratio becomes paramount. In future, with the development of 'personalized' data and medicine, it may become easier to compute this ratio for each man. At the moment, decisions are based on group data, always an approximation.

But, testosterone in animals has to have a much wider role on the body and the brain than simply activating reproduction,[50] if it is to be at all effective in promoting fertility and successful breeding, and this is the same in humans. A recurrent theme in this book is to emphasize the pervasive action of testosterone in all aspects of human life, as part of its central role in regulating reproductive success, the only definition of 'survival of the fittest'. A foretaste of this idea appeared in 1932, when Zuckerman[†] suggested that the persistence of primate social bonds, which he considered to be characteristic of this order, was the consequence of equally persistent sexual attraction, unlike non-primates in which sexual activity was periodic. Time has not been kind to this idea on a variety of grounds: other species do form prolonged social bonds, and as we have seen, sexual behaviour in some species of primates is also periodic, yet they retain their social cohesion. Ecological factors other than hormones play crucial roles in determining group sizes and persistence, and there are many other regulators of social bonding.[51] Nevertheless, It was an early exposition of the more general idea that the influence of sexuality and its hormonal control reaches well beyond the reproductive system. The fact that it does is the reason why testosterone, the prime mover of sexuality and fertility in males, has such a pervasive action on human lives and history.

But to be able to mate, a male needs more than motivation. He needs opportunity, and he lives in a competitive world. He has to measure himself against other males, and maybe fight for the right to reproduce. Testosterone equips him for this, as well. It turns him into a fighting machine.

[†] Solly Zuckerman (1904–1993) came to the UK from South Africa as a young man and was appointed anatomist to the London Zoo. He achieved early scientific and popular fame after publishing a book called *The Social Life of Monkeys and Apes* in 1932, based both on his observations of baboons in his home country, but also on a colony in the Zoo. But this group was abnormal (very few females) and the males fought persistently, often with fatal consequences. Although this did not represent a 'natural' condition, it told us something interesting about sexual competitiveness. During the Second World War, he became an expert on effects of bombing. He went on to become the UK government's chief scientific advisor, and Baron Zuckerman. A charismatic man, he seemed to know everyone, artists and scientists. For a biography, see John Peyton (2001), *Solly Zuckerman: A Scientist Out of the Ordinary*. Murray, London.

5

Testosterone and Aggression

What is aggression? In ordinary English usage it means an abridge-
ment of the rights of another, forcing him to surrender something he
owns or might otherwise have attained, either by a physical act or by
the threat of action. Biologists cannot improve on this definition...
except to specify that in the long term a loss to the victim is a real
loss only to the extent that it lowers genetic fitness... The essential
fact to bear in mind about aggression is that it is a mixture of very
different behavior patterns, serving very different functions.

E. O. Wilson (1975), *Sociobiology. The New Synthesis.*
The Belknap Press, Cambridge, MA

Violence, like all behavior and all disease, is multi-determined, i.e. it
is the product of the interaction between a multiplicity of biological, psychological and social causes, or variables (for example, male
sex hormones, child abuse, and relative poverty) each of which can
be shown to have the effect of increasing or decreasing the frequency and severity of violence....However, beneath all these
there are certain regularities and unities, one of which is that
shame is a necessary (but not a sufficient) cause of violence....

J. Gilligan (2001), *Preventing Violence.*
Thames and Hudson, New York

Anybody can become angry, that is easy; but to be angry with the
right person, and to the right degree, and at the right time, and for
the right purpose, and in the right way, that is not within everybody's power, that is not easy.

Aristotle, *The Art of Rhetoric.* Quoted by
E. Young, *New Scientist,* 9 February 2013

Testosterone may well make a male eager, even desperate, for sex. It may also make him fertile. But all that is not enough. Sex doesn't happen in a tranquil world. It's competitive: many males may prefer certain females; females have preferences too. A male needs to be competitive, aggressive—warding off rivals, fighting his corner. He also needs to be attractive, for although in many species, including humans, males are generally stronger than females, is it the females who have the final choice. Forced mating is practically unknown among non-human mammals, though interestingly it seems rather frequent in orang-utans.[52] We'll consider why it can be prevalent in humans later. So the males of most species have to see off rivals, and then persuade the female to mate with him. The two are strongly linked.

Male red deer live together peaceably in social groups for much of the year. But as the autumn breeding season approaches, this changes.[53] The groups break up, stags becoming highly aggressive towards each other, at the same time growing large, jagged antlers. They use these to fight, locking them together. This may seem picturesque to a human onlooker, but if these antlers should pierce the other stag's flanks terrible, even fatal, wounds result. This periodic aggressive, anti-social behaviour does not occur if the stags are castrated, but can be reinstated at other times of the year by giving them testosterone. They fight for only one reason: to gather a group of fertile hinds, and protect them from the sexual attentions of other stags. Red deer illustrate many of the complexities of male sexuality and aggression, and the reason why testosterone has such dramatic and widespread actions on both body and behaviour.

Red deer are not alone. Male rhinos charge each other; male giraffes flail each other with their necks; male hippos grasp each other's mouths. Elephants have periods of *musth* when they become highly aggressive, go on the rampage, and are sexually active. Zookeepers are very wary of *musth*: they know how dangerous it can be. Throughout the mammalian kingdom, males—mostly young males—battle each other for access to mates, or territory, which allows them to gather food and mates. And aggressive behaviour is not confined to mammals. Some of the most elaborate colourings and displays are seen in birds: the peacock being the prime example. All are dependent on testosterone. There are dozens of other striking examples of the power of testosterone over every aspect of a male's life during the time he has the opportunity to breed. Since the breeding season is, for many animals, limited, this expenditure of energy and, equally, exposure to risk, is also limited but intense. The strategy is to ensure that the risk of sex-related aggression results in sufficient fertile matings to generate enough young: 'enough' depending both on the species and on the conditions in which it lives, but is not likely to endanger the survival of sufficient numbers of males to maintain the population.

Take a look at a male hamadryas baboon. These baboons live in the African savannah, in large groups containing many males and females. The big males (much larger than the females) grow massive capes of hair at puberty. These act both as protection (it's harder to bite them through it) but also as a signal: here is a strong male (rather like shoulder pads in human males). The males of other primate species have similar, though not identical, features. The faces of some become brightly coloured (e.g. mandrills); others develop a bright blue scrotum (guenons), and sit with their legs apart. Baboons also grow massive canine teeth that often protrude even when their mouths are closed. They have a habit of yawning widely: which exposes these enormous weapons—for that is what they are—for all to see. These canines develop at puberty and are a common feature of the males of many primate species. They can do dreadful damage. The males most successful at defeating other males will mate with most females.

Human history shows clearly that a man's social status has had dramatic consequences for his reproductive ability. We know this because the Y chromosome, unlike others, is inherited intact from a man's father. This allows us to trace the inheritance of men; if they have the same Y chromosome, then they will have had the same ancestor at some point in their genealogical history.[54] Modern genetics has enabled detailed analysis of individual Y chromosomes. There are some surprising results. The presumed Y chromosome of Genghis Khan, a notable warlord who wielded absolute authority over his subjects, is found in around 8% of Central Asian men to this day, and he is said to have about 16 million descendants. There are other oligarchs who have left a similar genetic imprint on the modern world. King Solomon had 700 wives and quite a few concubines as well. The Incas formalized the relationship between social status and sex: aristocrats were allowed 50 wives, the heads of 100,000 men had 20, but those who commanded 10 men had only three.[55] Social status depends on many qualities, but the ability and/or willingness to fight successfully, or to dominate rivals, is a central one. Abraham Maslow* proposed that humans had a 'hierarchy of needs'. Food and water are basic, security and stability (e.g. an established home) is second, then comes 'love and belonging', and finally 'esteem', which means social success and status. Evidence for a social structure dependent on status is very ancient in human culture. Kings, chieftains, leaders, elders, all are prominent in the history and folklore of every nation, all peoples. This structure reflects the hierarchy of males. At its root is the tendency and ability of males to compete and, if necessary, to fight for their position in this hierarchy. In some cases, laws of heredity have been devised to bypass the need to fight for status. Among the rewards of high status is access to the more desirable females.

---

* Abraham Maslow (1908–1970) was an American psychologist who had considerable influence on 'normal' psychology (as apposed to psychopathology). He postulated that humans had a 'drive' for social dominance, hotly debated to this day.

It is important to recognize that aggression, unlike other behaviours important for survival and success, has no biological function on its own. Sex, eating, drinking, sheltering all directly influence whether a male, or his genes, will survive. Aggression acts as a conduit to these assets. It only makes biological sense if it occurs in order that a male improves his ability to survive, and this is generated by his ability to feed, mate, etc. So aggression has to be considered as part and parcel of these other behaviours. It is equally important to separate these 'ultimate' goals from the more 'proximate' causes or triggers for aggression itself. You think you eat for pleasure or to stave off the unpleasant pangs of hunger (the proximate causes), but you actually eat to provide your body with supplies of energy and materials for growth (the ultimate cause). Proximate causes ensure that ultimate ones are fulfilled. The fact that aggression is an adjunctive behaviour does not mean that it might not itself be rewarding. Does this mean that there is something called an aggressive 'drive'? If a man is deprived of food, he seeks progressively more urgently to eat the longer that deprivation lasts. The same goes for the urge to drink following prolonged water deprivation, or for sex following abstinence. Does this apply to aggression? Or is aggression a response to a demanding situation? It's an important question, because it has huge implications for how aggression is activated or controlled. Thomas Hobbes[56] ascribed aggression to three sources: competition, 'diffidence' (which we would call fear or defence), and glory (honour).[†] Note that the first two are apparently different from the third. Individual (or group) fitness would be directly affected by successful competition for resources; so would a successful defence of those resources against others. But achieving 'glory' reflects the tendency of males to preserve or enhance their social

---

[†] 'Let me not then die ingloriously and without a struggle, but let me first do some great thing that shall be told among men hereafter.' Homer, *The Iliad.*

order, and hence their overall competitiveness. It's a proximate cause. Konrad Lorenz recognized that aggression was an essential part of an animal's ability to survive in a hostile and competitive world, but he also thought that an aggressive 'instinct' or 'drive' built up until satisfied by a fight.[57] His views have been heavily criticised.[58] An alternative is that aggression is a product of a man's interaction with society, not a basic component of 'masculinity'. If this is true, then aggression is not an essential part of the action of testosterone on the brain, though it may activate aggression.[59] There is of course, an intermediate position: that aggression is a biologically important ingredient of male sexuality (and hence of testosterone), but one that is moderated by social control, circumstances, experience, and individual characteristics. That is the one promoted in this book.

The unique nature of aggression—that it is biologically part of some other behaviour—means that it has been unusually difficult to define, and even more difficult to measure. If we say that someone is 'aggressive' what do we mean? That he picks fights, or frequently uses abusive language? That he typically behaves in an aggressive manner towards, say, his workmates? But he may not do so towards his family. If we say a man's business methods are 'aggressive' this does not entail any risk of physical injury to anyone. Some men seek aggressive methods to further their ends: this is proactive aggression. Others are only highly aggressive when challenged by someone else: reactive aggression. The two are not the same, either biologically or physiologically. Is aggression in a sexual context the same as in competition for food, or social status, or as defence to an attack by another? Are anti-social behaviours and delinquency part of aggression? Similar problems face scientists who try to measure 'aggressiveness'. Females may be very aggressive in particular contexts: for example, in defence of their young. Males may defend their territory, but be non-aggressive outside it. Is aggression by one group towards another (e.g. war) the same as individual aggression (Chapter 8)? Psychologists, sociologists, and ethologists struggle to provide a unifying definition of aggression, or how to assess the aggressiveness of individuals.[60] Experimental scientists often confine their models to a particular context: such as one mouse invading the territory of another—an important but specific case of aggression. But information gained in this way may not apply to aggression in other circumstances. Quantifying aggression itself is another problem: it's quite difficult to say how much aggression is going on, rather than a simple 'there's a fight'. In truth, experimental studies on aggression have not proved very useful for understanding human aggression; for example, mice use their sense of smell during aggressive encounters rather differently from humans. Arguments about whether testosterone influences aggression are immediately faced with these difficulties; what sort of aggression are we talking about? Aggression is thus not a simple, unitary, behaviour; of course, other behaviours are not that simple either, but the complexity of aggression stands out (Fig. 11). The context in which it occurs is paramount, even though the physical manifestation (teeth, claws, sticks, or fists) may seem similar. Robert Hinde, a prominent ethologist, summarizes this viewpoint thus:

Individual aggression is often categorised into a number of types. For instance, one system distinguishes 'instrumental aggression,' deliberate and concerned primarily with obtaining an object of position or access to a desirable activity; 'emotional aggression,' hot-headed and angry; 'felonious aggression,' occurring in the course of a crime: and 'dissocial aggression,' regarded as appropriate by the reference group or gang, but not so regarded by outsiders. Such categories, though useful for some purposes, usually turn out to be less clear-cut than they might appear for an obvious reason: a variety of motivations may contribute to a single act, and they may be present in various strengths and combinations. The very fact that such categorization systems can only be partially satisfactory is in itself an indication of the motivational complexity of even apparently simple aggressive acts.

R. A. Hinde (1998), 'The Psychological Bases of War'.
American Diplomacy, vol. 3, electronic edition

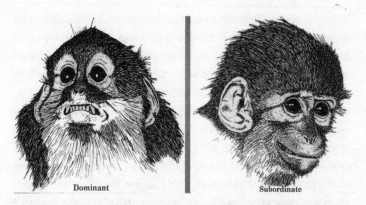

Fig. 11. Aggressive and submissive facial expressions in a species of small monkey (talapoin). Note exposure of the teeth and raised eyebrows in the dominant (aggressive) male. The submissive face may be the origin of the human smile.

Aggression is an excellent example of behaviour that carries a cost/benefit outcome. Nothing is gained simply by winning a fight. This doesn't mean that winning a fight is not rewarding in itself, or that aggression cannot be pleasurable for its own sake. But no individual increases his/her 'fitness' simply by defeating another. The payoff is the increased access to some desirable resource that overcoming another can bring. If a male chooses to fight for every female, then the chances that he will be seriously injured or killed increases. If he chooses to fight larger or better-equipped males, then he risks the same outcome. If he never competes, then he is unlikely to acquire a mate, let alone enough food, and so on, in an environment in which resources are scarce—the natural state of things, and one for which males of all species, including man, have evolved. So an obvious strategy would be for a male to weigh up the relative risk of a contest against the potential advantage of gaining some sort of asset (a mate, food, shelter, etc.). This, itself, is a high-risk strategy: if he makes a mistake, the consequences could be dire.

But there is another way, one based on social learning. Each male learns, either by trial or observation, whether another male is likely to triumph in an aggressive encounter. Many species, not only primates, have this capacity. It results in a social structure variously called social status, or dominance rank. So it is that, if you throw a peanut to a group of monkeys, it is likely that one (a large male) will move quite leisurely to pick it up. And the next one. Only when he is satiated, or there are too many peanuts for him to gather, do the others join in. Children feeding ducks in the park can observe the same process—the pecking order. Levi-Strauss records that, in some Australian tribes, the chief always has the first mouthful of any kill.[61] Competition is settled by this system of social learning; this reduces the risks to all members of the group. Repeated jousting is no longer necessary. But it does mean that assets are likely to be unevenly distributed between the males of any society. Achieving social dominance is an extremely significant contribution to survival and therefore 'fitness'.

A male's position in this hierarchy is not settled forever; evidence both from non-human primates and human society shows that a male's social position is constantly being monitored and tested by other males. Social ranking reduces overt aggression, but tension remains. Any change in the dominant male (including increasing age) may result in his being deposed by others. In non-human primates, the dominance rank between females, though it can exist, is nowhere so obvious as for males. Interestingly, in some primates, a female's rank may vary according to that of the male with whom she is currently consorting (there's an obvious human parallel situation).[62] A male monkey's rank may not depend solely on his physical qualities: if he forms a partnership with another male (one is tempted to call it a 'friendship'), then the two of them may, together, achieve a rank that is greater than either of them would on his own. Similar events happen in human society. Social status or ranking between males is a universal property of human society, and history shows us that it always has been. It is a powerful social mechanism for ensuring unequal, but arguably biologically sensible, distribution of resources without the need for constant fighting. One of the remarkable abilities of the human brain is to recognize this mechanism for what it is and, in some societies, try to take political and economic steps to moderate biologically derived inequality.

So how does testosterone fit in? We should bear in mind the complexity underlying the word 'aggression' outlined above. So we will not expect testosterone to play necessarily the same role in all the contexts in which aggression can occur. We should also remember that testosterone is only one component of the hormonal system, or indeed the physiological systems controlling the body, so anything it does will be in concert with other regulators and affected by them (this point is made in several places in this book, and needs to be). But what exactly are we trying to explain? Humans, like other animals, have to deal with a complex environment, which broadly separates into the physical world (climate, food, water, shelter, etc.) and the social one (other members of the same species in various guises—home group, rival groups, potential partners, potential rivals, and so on). As we have noted, the two overlap: for example, getting food or a mate may involve having to deal with competitors. Individuals have a range of options in most circumstances: they can choose to compete or retreat, to select opportunities (tactics) or individuals (targets) that promise the biggest payoff with the least risk or cost. The brain makes a distinction between ultimate objectives (e.g. getting food) and proximate ones (e.g. an appetite for risk or for fighting). So if the hypothalamus, the part of the brain that monitors the body's internal state (see Chapter 10) signals the need for food (ultimate), other parts of the brain signal ways to obtain food, and any risks or competitors that may lie in the way. One way of facilitating success may be to increase the appetite of the individual for competition, and the aggressive behaviour that goes with it (proximate). Are males more aggressive than females (again bearing in mind the caveat about what 'aggression' actually means) and can this be laid at the door of testosterone?

Go to any kindergarten and watch the little children playing. Anyone can spot the difference between boys and girls. The girls tend to sit quietly in a corner playing with dolls or drawing; the boys rush about waving sticks or pretend guns. They fight each other; girls seldom do (they tend to use verbal rather than physical aggression). It's called 'rough and tumble' play, and little male monkeys do it as well. Some sociologists have tried very hard to convince the rest of us that these differences are not innate, but the result of social learning or parental influence. Give a little boy some dolls and he will play happily with them as would a girl, they say. A great deal of evidence shows this view is too simple. The little boy is more likely to make two dolls fight each other! There is no denying that parents, siblings, and other significant people have huge influences on the way children behave (and play) and there are genetic and other variations between individual of either gender. There is therefore a range of tendencies for rough and tumble play across both sexes, as for any gender-differentiated behaviour. Despite all this, aggressive-type rough and tumble play behaviour is typically different in boys and girls. It seems to be a characteristic of the development of little boys. Is testosterone involved?

The evidence comes from observing little boys and girls who have CAH (congenital adrenal hyperplasia), the congenital syndrome in which they are exposed within the womb and subsequently (if not treated) to excess levels of testosterone (see Chapter 3). The little CAH boys' play is no different from normal; but little CAH girls play much more like boys than normal girls. Even at this early age, they show increased aggressiveness, as well the other aspects of male play behaviour. Are the parents treating these CAH girls like boys because they are, to a degree, masculinized both physically and behaviourally, and is this why they play in the way they do? The evidence shows otherwise,[63] but parents can influence play behaviour in CAH girls, just as they can in normal children.[64] The sociologists are not entirely wrong after all; it's yet another example of nature interacting with nurture. Individuals (genetic males) that are insensitive to their own testosterone (androgen insensitivity syndrome: AIS—see Chapter 3) play like girls, despite having as much testosterone as boys (and XY sex chromosomes). Notice that excess testosterone in boys does not result in 'super-males'—boys who show even more aggressive play than normal. So if we try to ascribe differences in aggression in normal boys to corresponding differences in their exposure to testosterone, our argument will sound weak. This is important, because there is no doubt that infant boys do differ in their play behaviour, and this may relate to similar differences in later life. If we want to ask why this should be, we may need to look elsewhere than at testosterone levels. It seems that normal levels of testosterone do everything that testosterone can do to aggressive-type play behaviour: more testosterone has no additional effect.

But aggressive-style play behaviour is just that: it's play. Unlike adult aggression, it seems to be done for its own sake (though fights may break out over toys, etc.). Nearly all young mammals play, so it has some important biological function. There is still argument over what exactly this might be, a topic outside the scope of this book.[‡] We need to consider whether aggressive play behaviour predicts aggressiveness in adults: here the evidence seems confusing. There is a history of aggressive play behaviour during childhood in many aggressive (e.g. criminal) adults (but also many adversities): but the converse is not so clear. An aggressive child does not necessarily grow up to be an aggressive adult. We need to be cautious about inferring roles for testosterone in adult aggression (of whatever type) from observing play behaviour.

[‡] Paul Martin and Patrick Bateson in their highly readable book *Play, Playfulness and Innovation* (Cambridge University Press, 2013) discuss the function of play in detail. They suggest that it forms the basis for later inventiveness and creativity.

What about the effects of exposure to testosterone in early life on aggression in animals and humans? Curiously, despite the remarkable effect on their sexual behaviour of giving neonatal female rats testosterone (Chapter 3), an equivalent result for aggression is not so obvious. However, female monkeys whose mothers were given testosterone are more aggressive than normal, and female hyenas, who are normally exposed to high levels of testosterone in the womb and are difficult to distinguish from males, are also highly aggressive. Human female twins whose other twin is a male are more aggressive than those in whom the twin is also female (it is supposed that the male twin exposes the female to some of his testosterone). As we have seen, CAH females, who are exposed to excess testosterone in the womb (see Chapter 3), show more male-type physical aggression than normal females but CAH males were similar to normal males.[65] There seems to be some slender evidence for testosterone during early life increasing aggressiveness in females, but no evidence that high levels in males can do likewise—an important theoretical point. Using the 2D:4D digit ratio (see Chapter 3) as an index of individual differences in exposure to foetal testosterone has also given equivocal results (but recall the caveats about this measure).

It is generally agreed that men (particularly young men) are more aggressive physically than women, though if the definition of aggression is extended to verbal assaults the debate becomes less certain.[§] Just over half of all violent crimes in the UK are committed by males aged 16–24, and over 80% involve male offenders.[66] But this also applies to victims of violence. You rarely see pub brawls involving anyone other than young males. Sexual assault is different: females are 10 times as likely as males to experience it. Testosterone is related to

[§] This does not mean that women cannot be very aggressive under some circumstances. For example, women guards in prison camps have been noted for their extreme ferocity.

this form of aggression, since one (but only one) requirement for this is adequate levels in the offending male. Rape is the most extreme form of sexual assault, and is discussed in more detail in Chapter 6. Attempts to answer the question: 'does testosterone control aggression in adult men?' have faced several problems. Giving men supranormal amounts of testosterone for anything other than brief periods presents ethical difficulties, though it has been done occasionally: aggression was not increased though 'roid rage' ('roid' = steroid), a supposed side-effect of illegal testosterone-like drugs, is said to occur (but this has been disputed). It may be linked to prior psychological problems in those who take these drugs as well as their excessive steroids. Illegal treatment with testosterone (or similar compounds) is a common feature of athletics and other competitive sports. It undoubtedly improves performance, which is one reason for its illegality. Muscle power is increased, though whether other testosterone-dependent attributes, such as heightened competiveness, also play a part is not clear. There are obvious difficulties in studying those who take unsupervised, clandestine and unknown amounts of steroid. Excessive testosterone can result in bizarre overdevelopment of muscles (and damage to other organs, including the liver) (Fig. 12).

PERCENT OF POSITIVE TESTS

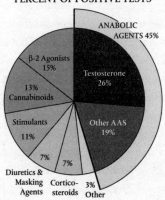

Fig. 12. Testosterone or associated steroids (AAS) are the commonest illegal drugs taken by sports athletes.

Most studies on how variations in testosterone are related to aggression are therefore correlations, and the old adage 'correlation is not causation' applies here as in other contexts. Then there is the problem of how to measure aggression: often by questionnaires, which can be misleading (which sort of aggression—proactive, reactive; which context?), or by asking subjects to play games with competitive or aggressive elements; but they are games, not real life. Sensitivity of individuals to their own testosterone depends, among other things, on variations in the genetic structure of the androgen receptor (see Chapter 2) and this, until recently, has seldom been measured. Testosterone levels alone may therefore be misleading. Finally, and crucially, testosterone levels in men vary from day to day, even moment to moment, and one important regulator is the amount of aggression they show or receive. Winning or losing, an intrinsic result of being competitive and aggressive, also can alter testosterone levels (see Chapter 7). So if there is a causal correlation, which way does it operate?

In 1981, D. R. Cherek, of Louisiana State University, invented a game. The subjects play against a computer, though they think this is a real person. They play for points, which are exchanged for real money at the end of the game. They have several buttons: one deducts points from their opponent, though they are not allowed to keep these points. Since this is damaging to their opponent without benefit to them, pressing this button is called 'aggression'. It's aggression for its own sake, since the biological function of aggression (a gain of some sort) is not fulfilled. Violent offenders press the button more than non-violent ones.[67] They seem to like inflicting harm. More interestingly, in one of the few studies in which testosterone was given to 'normal' adult men for several weeks, this behaviour increased. Even more transient increases in testosterone may have similar effects. In another game, subjects were allowed to play with a gun (aggression) or a board game (anodyne). Then their testosterone levels were measured and they were asked to add variable amounts of a hot sauce to a cup of water that would be given to another participant (this was classed as 'aggression'). Playing with the gun resulted in their adding more sauce, but only if their testosterone levels rose, which it did in some but not all individuals.[68] Other studies are similar: if testosterone rises during a competitive interaction, then the individual seems more willing to engage in a subsequent competitive or aggressive challenge.[69] Testosterone seems to make men like being aggressive, regardless of whether there is any benefit. We can easily relate this to the increased appetite for risk that is such an essential part of competitive reproduction. This is a classical 'proximate' cause of a behaviour: the 'ultimate' one is that males with this characteristic are more likely to take risks to win a mate, etc. A more nuanced view of the action of testosterone is that it stimulates ambition for higher social status, but this can be achieved using a variety of strategies. This may include aggression, but also generosity etc. to others in suitable contexts. So testosterone may set the target, but other constraints determine strategy. This is similar to the complex process of attaining sexual objectives.[70]

In Chapter 3, we described how the presence of a Y chromosome makes testes, which in turn make a man. But that may not be all the Y chromosome does. There is a rather rare condition in which, because of abnormal cell division during development, a male gets two Y chromosomes instead of the normal one. So he is XYY. You might think he would be overendowed, yet boys with this condition have normal levels of testosterone, and a normal-looking puberty. But in the 1970s, when techniques for assessing the presence of two Ys became available, it was noticed that XYYs were more likely to be convicted of crimes than was expected (Fig. 13A & B). Great excitement followed: had we discovered a gene on the Y chromosome for (male) criminality? Was the Y chromosome responsible for the statistical excess of (young) males in crime, particularly violent crime? Subsequent research has given somewhat mixed results, but XYY boys do seem more aggressive, impulsive, and delinquent than comparison

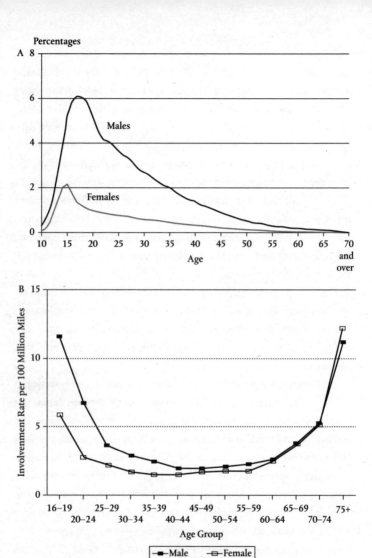

Fig. 13. (A) Criminal offenders as a percentage of the population by age and gender (Office for National Statistics, UK). Note the excess of young males, though young women are also more likely to offend than older ones. (B) Risks of a car accident by age and gender (USA). Note that while young men are more likely than young girls to have an accident, rates rise in older age in both sexes.

populations of XY. They are also taller (the Y chromosome may contain genes that regulate height), and may be less intelligent: all factors that could contribute to this difference. There are those who think that genes on the Y chromosome may contribute to aggressive behaviour independently of testosterone.[71]

Now is the moment to recall once again, if briefly, that testosterone does not act alone on aggression, any more than in any other context. While we will consider how testosterone acts on the brain in more detail later (Chapter 10), here I should mention several neurochemicals that have a role in aggression. The first is serotonin. Low serotonin levels in the brain have been associated with increased tendency for aggression, particularly reactive or 'impulsive' aggression. Consistent with this, variations in genes associated with the way that the brain handles serotonin have also been implicated. For example, serotonin, like all chemical transmitters in the brain, acts on receptors. In the case of serotonin, there are at least 14 different ones, as well as another type that sucks up serotonin from the synapse after it is released, and so limits its duration of action. One of the serotonin receptors (type 1B) has been particularly associated with aggression, and drugs to act on it have been suggested as therapy for pathological aggressive states. But serotonin is widely distributed in the brain and influences, as one might therefore expect, many functions including eating, sexuality, anxiety, and many other behaviours. One of them is 'impulsivity', the tendency to react quickly to any situation without much restraint, and associated with lower levels of serotonin activity. This may account for an increased likelihood for reactive aggression in people with certain genetic variants of serotonin receptors. But impulsivity is unlikely to be restricted to aggression.

Another gene associated with serotonin is called MAO-A.** MAO-A is an enzyme that breaks down serotonin, but also other related

---

** Stands for monoamine oxidase, type A.

neurochemicals such as noradrenaline (norepinephrine in the USA) and dopamine. Serotonin belongs to a chemical 'family' (the 'monoamines') and its activity cannot really be fully understood alone. There are two forms of the MAO-A gene: high or low activity. Childhood adversity is particularly liable to result in an aggressive adult in males with the low form.[72] another example of the way genes can interact with the early social environment: intriguingly, the opposite may hold for girls. This may seem paradoxical, since serotonin would be increased by low MAO-A. But the explanation may be this: serotonin acts on the serotonin neuron itself as well as the next one in the chain. It suppresses the activity of its own neuron: thus more will lead to less serotonin being released at the other end. Another explanation is that a brain with this MAO-A variant develops differently, and this accounts for impulsive or aggressive behaviour in adulthood. Serotonin is known to influence the way the brain grows. Whether testosterone could be responsible for this sex difference has not been explored. MAO-A has been called the 'warrior gene' because those with the 'low' variant may be more aggressive (this also applies to mice). This is a misnomer, since it implies some specific relation between this gene and aggression. But the systems on which the enzyme MAO-A acts have widespread roles in many behaviours, so aggression is only one result of the 'low' genetic variant.

Vasopressin is a very different chemical. Also called 'anti-diuretic hormone', it was first identified as an important pituitary hormone that regulates how much water the kidney allows to pass into the urine. So in hot, dry, conditions, urine becomes more concentrated because the pituitary secretes more vasopressin (a small peptide). People who lack vasopressin pass huge quantities of urine (the condition is called 'diabetes insipidus', and is quite different from the more familiar diabetes mellitus). But it has other functions. One is to play a part in the secretion of cortisol, the stress hormone, from the adrenal glands (it regulates the production of ACTH (adrenocorticotropic hormone), the large peptide controlling the adrenal glands, from the pituitary). Another, still not fully established, may be to accentuate aggression. Tiny infusions of vasopressin into the brain increased aggressive behaviour in rats. Strikingly, a part of the brain associated with aggression (but also other behaviours) has a lot more vasopressin in male rats than females. It gets better: castration depletes vasopressin in this part of the brain, whereas giving testosterone repletes it. Then the story stumbles: not all species have so much vasopressin in this area, nor do they show the effects of testosterone. But it may still be relevant to man: giving vasopressin to males increases a hostile facial response to unfamiliar faces, and makes the expression on those faces appear to be less friendly. Interestingly, the effects of vasopressin in women are quite different: it encourages friendliness.[73]

There are other hormones too that may be important for aggression, and they interact with testosterone. The hormone cortisol (corticosterone in rats) is elevated in stress, and may encourage aggression by increasing arousal in situations that include post-traumatic stress disorder, etc.; low levels of cortisol might encourage different types of aggression, such as those seen in childhood conduct disorders. Social dominance is associated with higher levels of testosterone (see earlier) but this may not occur in those who also have high cortisol levels (e.g. are also stressed).[74]

It is not surprising that we still know so little detail about aggression. It is so easy to recognise aggressive behaviour that this has disguised its true complexity—the context or circumstances in which it occurs. Factors that regulate it, including testosterone, which undoubtedly has a major role, can only be understood in the light of these contexts. We will discuss the part played by the brain in aggressive behaviour in a later chapter (Chapter 10), but here we only have to note that our incomplete understanding of the brain adds another dimension of uncertainty. What we can conclude is that the propensity to behave aggressively is an ancient and necessary quality of humans—particularly males—reaching way back into our history. In order to behave aggressively, males have to be prepared to take the risks of conflict, and even like to fight. Their brain makes men like what is good for them in a biological context. So they like to eat, to drink if they are water-deprived, and have sex. We can add aggression to the list. Life can still be a struggle, but it was perhaps even more precarious in primeval times. High levels of aggressive behaviour during competition for resources must have been common, as they still are in certain human societies. Conflict between societies—that is, war—is discussed in more detail later (Chapter 8), but the overlap and distinction between aggression between social groups and that within them, while having similar biological purposes (improved access to scarce resources) must be made clear here. As we will see, the actions of testosterone have relevance to both types of aggression, even though there are differences in the way they are manifest.

What has evolved are increasingly complex (and effective) ways of controlling or limiting aggressive behaviour within societies. What may have been appropriate and even necessary levels of aggression in earlier times have now become maladaptive. For example, it is said that homicide has fallen by about tenfold since the thirteenth century, though there are periodic increases.[75] Men have inherited a behaviour which has been essential from the earliest times, but can be counter-productive as the organization and function of society changes, though its importance in the contemporary world should not be underestimated. There are continued calls in Western societies for improved ways of reducing such aggression as still occurs. Upbringing, social mores, laws, and customs are all designed to regulate and channel aggression (but not abolish it). These have nothing to do with testosterone, whose primeval role in aggression continues, but with those parts of the human brain that are responsible for devising and applying such controls. These regions are intimately linked with being a human being, so that, while all other species have brakes and controls on aggression, in man these reach complexities and variations not seen in any other species. Similar social controls apply to other actions of testosterone, such as those on sexuality, with which aggression is closely linked. Chapter 6 considers this more closely: here we can simply point out the considerable national (cultural) variations in levels of violence. For example, the murder rate per 100,000 in the UK is 1.2, in the USA 4.8, and in Colombia 31.4.[76] Most murders are carried out by males. Assuming no dramatic genetic differences between these societies, it is clear that social control remains very important even in the modern world. Testosterone is simply so powerful that there have to be means of controlling its action.

6

Controlling Testosterone

Nature has placed mankind under the governance of two sovereign masters, pain and pleasure. It is for them alone to point out what we ought to do, as well as to determine what we shall do. On the one hand the standard of right and wrong, on the other the chain of causes and effects, are fastened to their throne.

> Jeremy Bentham (1780), *An Introduction to the Principles of Morals and Legislation*

She had been very pretty when she was young... Another of her favourite stories was of the day she had danced with a real lord... before the evening was over he had whispered in her ear that she was the prettiest girl in the country, and she cherished the compliment all her life. There were no further developments. My Lord was My Lord, and Hannah Pollard was Hannah Pollard, a poor girl, but the daughter of decent parents. No further developments were possible in real life, though such affairs ended differently in novelettes. Perhaps that was why she enjoyed reading them.

> Flora Thompson (1939), *Lark Rise to Candleford*. Oxford University Press, Oxford

Reason is always a kind of brute force; those who appeal to the head rather than the heart, however pallid and polite, are necessarily men of violence. We speak of 'touching' a man's heart, but we can do nothing to his head but hit it.

> G. K. Chesterton (1958), 'Charles II'. In: *Essays and Poems*. W. Sheed (ed.). Penguin Books, Harmondsworth

Voluntary, the third, or intellective [moving sense] which commands the other two [appetites] in men, and is a curb unto them, or at least should be, but for the most part is captivated and over-ruled by them: and men are led like beasts by sense, giving rein to their concupiscence and several lusts. For by this appetite the soul is led or inclined to follow that good which the senses shall approve, or avoid that which they hold evil.

Robert Burton (1621), *The Anatomy of Melancholy*

The optimal strategy for any male seems obvious at first sight. He should mate with as many females as there are around, selecting the fertile ones if at all possible, without restraint or hindrance. Whether or not he invests in the care or protection of any subsequent offspring would depend on how certain he could be that they were his. In this way he ensures the maximum spread of his genes, and therefore his biological 'fitness'.[77]

In fact, no mammalian species allow this to happen. Thomas Hobbes' postulated 'state of nature'[78] (a situation in which there is no governance, social organization, or laws) has never existed in animals, or in man. Some of the clearest examples come from non-human primates. Adult male baboons are large animals with formidable teeth; their canines can rip another one's hide to shreds in minutes. They can run very fast, so a miscreant has little chance of escaping. Despite their daunting appearance, they tolerate young males, who may play around them, even climbing onto their backs, or pulling their tails. This period of domestic harmony comes to an abrupt end as soon as the little males enter puberty. Then the scene changes dramatically. Far from being tolerated, the young males are driven from the group to form disconsolate bachelor bands, roaming the surrounding landscape with other exiles. If they are lucky, they can gradually assimilate with other, neighbouring groups; there they may gradually insinuate themselves, and eventually fight their way to positions where they, too, can mate and drive out competing males. But there are those who are not successful, and live out their shortened, brutish lives outside any group. Even male chimpanzees, who live in large groups containing many males, and in which sexual activity is widespread (but not indiscriminate or promiscuous), are constrained by the other members of their group.[79]

The pubertal testosterone-dependent signals that release such antagonistic responses to the young males from the elders of their home group are not clear. There could be many: puberty, characterized by rapidly increasing secretion of testosterone from the young male's testes, can alter his appearance (including the colour of his face or his genitals), his smell (testosterone in some monkeys, as in humans, has powerful actions on glands in the skin), or his behaviour (his response to other males, or adult females). Something about this young male changes as the consequence of surging levels of testosterone, and this releases severely antagonistic responses from the resident adult males. This has two consequences: it removes potential competitors (or even the chance of furtive matings) from the adult male's immediate environment, thus increasing his potential fitness (any young are more likely to be his). But it also ensures that genetic variation is maintained. If the young male were to remain in his group, then the chances are that he might mate with a relative—perhaps quite a close one. Monkeys, like humans, carry recessive genes that are potentially harmful. If an individual carries one recessive (deleterious) gene but a second ('normal') dominant one, then the recessive gene will have little or no effect, and there will be no physical or functional deleterious consequences that might otherwise occur from the action of the recessive gene. Since close relatives are likely to carry similar genes, then mating between close relatives increases the likelihood that two recessive genes will occur together. In this case, the recessive genes will be active, with possibly dire consequences.[80] So driving out the young male to another group reduces this possibility, and maintains what geneticists call heterozygous fitness (meaning the combination of two different gene types: possession of only one type—either two dominant or two recessive—is called 'homozygous'). Humans have long recognized this, without any recourse to genetic knowledge; hence the widespread ban in many societies on brother–sister mating and, in some, on marriage between first cousins. Studies on human societies that allow cousins or other close relatives to marry show a hugely increased incidence of disabled offspring from such unions.[81] Sexuality between siblings may also be controlled psychologically. Children reared together (either by the same mother or, for example, in a kibbutz) show a much-reduced likelihood of marrying or having sexual relations; they are less likely to find each other sexually attractive. So as well as social aversion to incest, there may be biological barriers discouraging homozygous mating based on early experience.[82]

Driving out potential rivals before they can compete is a dramatic way of controlling sexual behaviour in monkeys and apes, but it is not the only way. Nearly every non-human primate species has a social mechanism for regulating males' sexual activity. It is striking that this varies so much between different species. Some species of baboons, for example, live in groups that contain several adult, fertile males. In this case, a particular control system operates. The males form a 'dominance' hierarchy (see also Chapter 5). Essentially, this means that each male learns the likelihood of the outcome should he challenge another male to a fight. Most likely, this is based on a previous encounter, or perhaps an assessment of the relative size or strength of a potential opponent. The male at the top of the hierarchy has privileged access to a variety of desirable and restricted assets. These include not only food and shelter, but also females. Some males become dominant by forming coalitions with other males, even though, individually, they might be defeated. Lower-ranking males are not only inhibited from mating by the likelihood of being attacked by the superior male, they may also show less willingness to mate even given the opportunity. Being stressed, and living in fear, increases chemicals in the males' brain that inhibit sexuality.* But being sub-ordinate may also lower their testosterone (Chapter 7), another way in which their sexuality is reduced. All these restrictions regulate sexuality in monkey groups containing several males.

But if this leads you to believe that sexuality in such groups is controlled entirely by males, you will be surprised to learn that even in baboons or macaques,[†] species in which the male is very much larger and more powerful than females (who also lack those savage canine teeth), the females have considerable say in selecting their partner. They don't necessarily prefer the most dominant male, though—interestingly for what follows—being dominant does seem to confer added sexual attractiveness on a male. So rampant male sexuality is also controlled by female choice.

Other primate species adopt different strategies to limit testoster-one-driven sexuality. Some form single male–female bonds that may last for years.[‡] This restricts male sexuality to one female, and excludes those unable to find a consort. In other species a male will gather a 'harem' of females around him,[§] which he defends against any other male, and this limits their sexuality while enhancing his own. Of course, from time to time furtive matings with other males occur. Interestingly, though polygyny (one male and several females) is common in non-human primates, polyandry (i.e. one female and several males) is very rare and occasional and not typical of any species. Comparative studies on a wide variety of human societies show a similar pattern. All these are examples of the ways that social organization restricts and channels male sexual behaviour, so that despite the powerful urges initiated by testosterone, these are corralled by the circumstances in which the animals live. In some species this seems

---

* High levels of the stress-related hormone cortisol tend to inhibit sexuality.
† The rhesus monkey is a common macaque.
‡ Marmosets and gibbons, for example, form long-lasting male–female bonds.
§ For example, hamadryad baboons.

less strict than others: bonobos, for example, who live in groups containing many males and females, mate rather promiscuously.

Before we leave the non-human primates, we should recall that similar mechanisms, though not so general or obvious, do exist in females. The regulation of mating by social rank is, in general, less evident in females, and they do not drive rivals out of the group. But in marmosets, for example, a daughter will not become fertile until either her mother dies or she leaves the group: her reproductive system is inhibited for so long as she lives with her family. Once she is removed from them, her ovaries are activated. This not only prevents incest, it also ensures that any young carry the genes of the male's partner. In those rare species in which females outrank males, despite their smaller size (and there are one or two**), the more dominant females may mate more frequently and have more young, but this selective action on reproduction is less pronounced than in the dominant males of other species.

This panoply of sexual tactics in non-human primates, even though it has as its objective a more consistent strategy—restricting male sexuality—presents us with problems if we are to consider what, if anything, our common primate heritage bequeaths us as the means by which human sexuality is restrained or directed. There is no doubt that the various social mechanisms that have been observed in non-human primates strike a resonance in man. Humans have a wide range of socio-sexual constructions both across continents and time, but we are a single species rather than the numerous non-human primate ones. Support for any one of the wide variety of human social structures could be found in that offered by other primates. It's clear that we cannot simply assume that we bring into our world some supposed ancestral 'primate' pattern. But is the very variety of sexual

---

** The talapoin monkey, a small species that lives in large groups on the river banks of west Africa, is one.

control exerted by other primates some sort of primal reason why humans have, and do, adopt such a large range of solutions between societies and across time to a common problem: containing masculine testosterone-driven promiscuity?

The powerful controls over male sexuality exerted by religion, laws, ethics, and social customs are tribute to the variety and effectiveness of these restraints. As in many other realms of human activity, the human brain brings to this area a huge and impressive set of complex mechanisms that go way beyond the social controls existing in other species, effective as they may be, though there are obvious similarities: for example, the existence of social dominance between males as a method of control. Unlike other species, the exact nature of these controls in humans, and the way they are applied, has differed over the years, and continues to differ between cultures. Compare, for example, contemporary sexual mores between the USA and, say, Saudi Arabia, or, historically, between the UK in Victorian and modern times.

But there are some general principles. The objective is always the same: to limit and channel male sexuality in ways that a particular society considers appropriate. Much of this is concerned with ensuring that females mate only with the prescribed males, and that the distribution of sexual activity conforms to the desired structure of that society. Thus the religious emphasis on prohibiting incest, or preventing sex before marriage; or the limitations on sexual interaction between different classes or castes within a society, or different religions, or the legal constraints on multiple wives (in some cultures). Many of these are ways of protecting the status quo, and discouraging social change. These prohibitions are exerted not only by external agencies, that is by penalizing those who break the rules—but by instilling in-built prohibitions: social inhibitions, or a conscience—during the process of upbringing.[83]

There are thus two major objectives of controlling sexuality—restraining the behaviour of males, but also regulating sexual competition between them. Having acquired a female, males adopt a variety of tactics to ensure that they keep them. Many methods have been used: chastity belts, guards on harems, drastic social and physical punishments on erring wives (fewer on husbands, usually[84]). Men have been particularly successful in persuading women that some of these procedures are in their interest. A good example from contemporary society is wearing a niqab (or burqua). This is really a way of concealing a woman's features or her body, and hence limiting the chances of her being attractive to other males. It is a male reproductive stratagem, but now inculcated into the religious and social mores of women. None of these complex social mechanisms owes anything to testosterone, but to those parts of the human brain concerned with higher levels of cognitive function: the huge development of the cerebral cortex that is the prominent feature of humans. Damage to the frontal lobes (particularly well-developed in humans), for example, can result in uninhibited sexuality: social controls are no longer effective (see Chapter 10 for a more extended discussion). The inventive capacity of the human brain is as evident in the social regulation of sexuality as it is in all other human activity. But the fact that such controls are deemed essential in all societies owes nearly everything to testosterone, for, as is abundantly clear from the rest of this book, it is this hormone that generates the powerful motivational and emotional states in men that give rise to the competitive nature of male sexuality and all the consequences this has for social control.

In humans, as in many other species, sexual selection or preference by females exerts a major control on male sexuality. We also know that there are circumstances in which this control breaks down: we call it rape. Rape represents the breakdown of the normal social controls on male sexuality. There is a huge literature on rape (we are only considering heterosexual rape here), its incidence or prevention, the forms it can take, the social and legal attempts to limit or recognize it, and the way these have changed over time or differ between cultures.[85] Here we need to note two facts: the first is that rape undoubtedly occurs; the second (less often recognized) is that, since rape is a form of aggression,[86] studying it presents many of the difficulties associated with defining and elucidating aggression (see Chapter 5). So it is that rape during war, or in prisons (the latter mostly man on man), or by acquaintances (e.g. 'date rape'), or within marriage will not necessarily be similar in all respects, even though they share the common feature of forcible sexual intercourse. But just as two fights may look the same to an observer, but have significant differences according to the context in which they occurred, so can rape.[87] Though rape is a considerable social, psychological, and legal problem, the huge majority of sexual encounters in humans are consensual and, in most circumstances, if a female does not wish for such an encounter, the male accepts this and it will not occur. That is, sexual selection by females is an effective, but not infallible, control on male sexuality in humans, as in other species. So what really concerns us here is not why rape happens, but why it mostly does not happen. But first we have to consider the role played by testosterone.

Whether or not rape is made more probable by excessive amounts of testosterone is debated. Some find that levels are no higher in rapists (but recall the caveats about the heterogeneous nature of rape), but others disagree.[88] Interestingly some rapists, at least, might be unusually sensitive to their own testosterone since there seems to be an excess of those with the 'shorter' (more sensitive) form of the androgen receptor variant[89] (see Chapter 2 for more details on this receptor). Variants in other genes associated with a tendency to aggressive behaviour are also possible candidates (e.g. those associated with serotonin, see earlier), but evidence is lacking. Those taking illegal androgens (anabolic steroids) are not, it seems, more likely to commit a rape. Various jurisdictions have, from time to time, instituted treatments for persistent rapists that involve attempts to reduce the activity of testosterone. Castration is one. Drugs that are 'anti-andro-gens' are another: these act by blocking the access of testosterone to the androgen receptor, and are regularly used to treat prostate cancer. Treatments that lower the pituitary's secretion of the hormones that regulate the testis, the gonadotrophins, are a third. None of them has gained general acceptance, either because they are ethically question-able, or because they depend on the subjects reliably taking the drug, but more pertinently because (as the Romans knew) post-pubertal castration or reduction in testosterone is not always very effective in lowering sexual activity in adult men. Furthermore, rape may be more of a dysfunctional aggressive event than a sexual one in some cases. Nevertheless, it is clear that, as pubertal testosterone is responsible for the emergence of adult sexual behaviour as well as its associated tendency for aggression, the surge of testosterone at and after puberty is necessary (but most certainly not sufficient) for a man to commit rape.

Forcible copulation is rare, even unknown, in most mammalian species other than humans, though one can always argue that obser-vations have been insufficient. Two exceptions are chimpanzees and orang-utans; since these are also large-brained animals, it may be that a larger brain enables more flexible approaches to sexual coercion in primates. But there may be other explanations. Females of most species that are not 'oestrous' (in heat) are both sexually unreceptive but also unattractive; they may also fail to adopt the posture necessary for a male to mate. This eliminates infertile (biologically inappropri-ate) mating. This is not the case in humans. The males of most species use more or less elaborate courtship rituals to persuade a particular female to mate. These, it is supposed, give her an indication of his biological 'fitness'. Males also, as we have seen, compete with each other for access to females, and use a variety of reproductive and behavioural strategies to retain possession of females or limit the access of other males. Violence is generally between males, not between males and females. Courtship can be persistent and vigorous, and the boundaries separating the transition to coercion and thence to force have been much debated among those familiar with chimpanzee behaviour as well, of course, as in humans.

A major difficulty with assessing how far control by females over male sexuality has lapsed in humans is the haziness surrounding statistics on rape and sexual violence. It is said that over 75% of all rapes may go unreported in the UK.[90] In other countries, this figure may be even higher. If this is true, then conclusions based on the incidence of rape are meaningless. This needs to be borne in mind over the next few sentences. Furthermore, there is continued debate over exactly what constitutes rape. Nevertheless, it seems that heterosexual rape probably occurs in humans more frequently than in any other recorded species, with the possible exception of chimps and orangutans. Even within Europe, the incidence seems to vary greatly. For example, the UK rate is around 29 per 100,000, whereas in the Netherlands it is about nine.[91] It is said that around one in six women in the USA experience attempted or actual rape during their lifetime[92] (but there are disagreements over the definition of 'attempted' rape). If these differences between societies are reliable (recall the unreliability of statistics on rape), they suggest that the powerful controls over male sexuality exerted by social, religious, judicial, and educational constraints also vary greatly. We lack more detailed understanding of this significant fact, though this might lead to improved protection for women of all societies. Interestingly, homicide—another aspect of illegal aggression—is no different between the UK and the Netherlands. Other countries show similar huge differences: in Papua New Guinea, fully 41% of men admit to having raped, whereas in Sri Lanka it is only 6%.[93] If these figures do not simply reflect differences in reporting, then they are another vivid illustration of how social control and customs vary between different cultures.[94] In the UK, about three-quarters of all rapes are committed by men under the age of 40, and other countries are similar. There are many other relevant factors—poverty, family history of violence, mental illness, low educational achievement, and so on—that seem to predispose a man to committing rape. These are all individual characteristics. But one of the contexts that consistently features in any account of rape is war.

We consider the role of testosterone in the phenomenon of war itself in more detail in Chapter 8. Here we are concerned with the fact that rape is commonplace in most, if not all, recorded wars. This is irrespective of the nationality, and therefore culture, of the soldiers concerned. It seems to have been almost a constant feature of military action throughout history. Does this tell us anything useful about why rape is less common under more normal conditions?

Wars are actually fought by young men, though older men declare wars and direct them. As discussed in other chapters (Chapters 5 and 8), young men are liable to be territorial, competitive, tend to form closely knit groups (e.g. street gangs), and easily become dedicated to a 'cause'. That's why terrorists are usually young men. There are arguments for ascribing all these characteristics to the direct or indirect actions of testosterone (Chapters 5 and 8). Rape in war is not only common, it is often committed by young men who would never think of perpetrating such a crime in civilian life. Why, then, does loss of the usual social and personal control of this testosterone-driven behaviour occur? First, there is testosterone itself. Levels in soldiers in the midst of a battle, fear and stress abounding, can fall to very low levels (Chapter 7).[††] In the more relaxed and triumphant atmosphere of victory, they will rapidly climb. This steep and sudden surge may be greater than in most normal occurrences, and might explain the equal

---

[††] Increases in other stress-related chemicals in the brain may contribute to diminished sexual activity in soldiers during active service. One is called â-endorphin, a substance related to opiates such as morphine. Morphine addicts are well known to retain little interest in sex.

surge of sexual motivation in victorious soldiers. But it does not explain the loss of control.

Enemy civilians may be dehumanized, particularly if they are ethnically different, thus reducing socially learned restraints. The social environment of a war zone is very different from that in which the soldier was raised; normal rules do not apply. The fact that a soldier may observe other members of his tightly knit group commit rape may both legitimize and stimulate his own behaviour. Excess alcohol may reduce the ability of a soldier to control his actions (similar to that seen in many cases of civilian rape). Rape may be used as part of the process of humiliating a defeated foe. All these factors imply that those parts of the brain, for example the frontal lobes (see Chapter 10), that are associated with social restraint, are rendered temporarily dysfunctional during the highly emotional and exuberant circumstances of victory. Add to this the fact that the frontal lobes of young men have still not achieved their final mature state, and are thus the more easily overridden. But more primeval forces may be at work. Overcoming a rival group has, as one reward, access to the group's females (this occurs also in non-human primates). Fertilizing your enemies' females makes biological sense, even though it may be socially repugnant. But it is the total breakdown of two normal regulations, female choice and male constraint, in the presence of testosterone-driven behaviour that typifies rape in war. How far this applies to other contexts is still discussed though, as already mentioned, it is important to take context into account. Rape is aggression and, as Chapter 5 relates, aggression is not a simple behaviour, with a single set of attributes. Attempts to understand why rape is so common in war will not necessarily apply unconditionally to rape occurring under other circumstances.

The true incidence of rape in war is no easier to estimate than in civil life, but some of the statistics are horrific. For example, it is said that around 100,000 German women were raped after the fall of

Berlin in 1945.[95] Many other wars have similar histories.[96] And there is evidence for a second primeval response, but this time in the male partners of those women who were raped. Despite knowing that these women had no say in their misfortune, or even might have been killed had they not acquiesced, many of these males were unable to accept this fact, and abandoned them. This reflects the biological fact that any children of a raped woman may not be theirs, though this might not have been the apparent reason why they had been rejected.

The control exerted by a current social system over male sexuality can break down in other ways. During the immediate aftermath of the Second World War, when much of Europe was ruined, and previous social structures had largely broken down, it was noticeable that the women of countries that had been occupied fraternized eagerly with soldiers of the liberating armies, in defiance of the pre-existing social order.[97] Access to scarce resources, like food or luxuries, was undoubtedly a factor. But it was not the only one: women often found the new arrivals more attractive than their own compatriots, diminished by defeat and offering little. And they were expressing a newly found sense of independence which would, in the coming years, result in the growth of the women's liberation movement, one of the most significant socio-political events of the twentieth century. Native males reacted strongly, as their testosterone-driven temperaments would predict, and such fraternizing women were often severely punished. An ancient imperative still persisted: it is not in the biological interest of males to lose control over sexual behaviour in their own societies.

The fundamental biological need to regulate male sexuality is as prominent in humans as in other species. We have not invented the need for social control. What the human brain has done in this context, as in so many others, is to devise unique, complex and varied ways of carrying out an ancient function, essential long before the advent of mankind: to limit indiscriminate testosterone-driven male sexual behaviour. There is a balance to be made between the severity of such controls and the ability of males to seek out, court and compete for females—a general biological need. History and contemporary sociology show us that the way this balance has been poised has varied hugely. In some societies, young people make their own way through the competitive, controlled world of sexual coupling: in others, parents decide for them, decisions often restricted by social caste, status, or money: so the competitive element is carried by them, rather than their children. This variety—there are many others—is another unique feature of the interaction between the basic, ancient biology of testosterone and its regulation by complex human behaviour, enabled by the evolution of the human brain.

Reproductive fitness and choice implies inequality: there are winners and losers. Testosterone not only influences who will be which; it also responds to the outcome. This two-way interaction colours how testosterone relates to the competitive world that surrounds men in all walks of life, in all societies—including some surprisingly modern situations.

7

Winning, Losing, and Making Money

Serious sport has nothing to do with fair play. It is bound up with hatred, jealousy, boastfulness, disregard of all rules and sadistic pleasure in witnessing violence. In other words, it is war minus the shooting.   George Orwell (1945), *The Sporting Spirit*. Tribune, London

Billions of shares are traded every day...Most of the buyers and sellers know they have the same information; they exchange stocks because they have different opinions...The puzzle is why buyers and sellers alike think that the current price is wrong...For most of them, that belief is an illusion.

Daniel Kahneman (2011), *Thinking,
Fast and Slow*. Macmillan, New York.

I'm tired of Love: I'm still more tired of Rhyme.
But Money gives me pleasure all the time.

Hilaire Belloc (1870–1953), 'Fatigue'

Testosterone levels do not stay the same for years and years; they vary all the time. They have a daily rhythm (highest in the morning) but, more importantly, they reflect what goes on in a man's life, particularly winning. Having sex increases testosterone levels.[98] Testosterone is also increased by talking to an attractive woman, or watching pornographic films. But levels are particularly sensitive to competition. Not only does testosterone encourage males to compete, its levels also respond to whether they win or lose. Increased testosterone after sex might be regarded as one example of a 'winner' effect, since having sex implies success in the competitive world of sexual selection. Whether increasing levels in this way has any functional significance is uncertain: one interpretation is that it ensures that a male with access to females has enough testosterone to enable him to be fertile. Conversely, a male without any opportunity for sex may suffer lowered testosterone, thus accentuating his uncompetitive state. Much of the evidence for the 'winner' effect on testosterone comes from studying sports, which you could say is not real life.[99] But many of those who compete in games take them extremely seriously, and whether they win or not matters to them hugely, so that sport can be a proxy for real life. Of course, for those who are professional sportsmen, it *is* their real life: if they lose too often, they risk their livelihood.

The attraction of studying sport is that, unlike the rest of life, it is a standard, repeatable, and clearly defined competitive event with a clear outcome, defined by the rules of the game. The variety and turbulence of existence makes it difficult to design a scientific study in ordinary life. But even sport presents problems: one is that it usually involves considerable physical effort, and this alone may alter testosterone (as well as many other hormones, including the stress hormone cortisol). So it may be difficult to detect changes that depend only on the outcome, rather than the process. Another is that sportsmen usually train in some way, and the process of getting fit can affect testosterone (and other hormones). Some of these difficulties are avoided by studying competitive games that require mental, rather than physical, exertion—for example, chess. But sport can be so entrancing and compelling that it takes on elegiac and emblematic properties:

> ...Though basketball was his sport, Rabbit remembers the grandeur of all that grass, the excited perilous feeling when a high fly was hoisted your

way, the homing-in on the expanding dot, the leathery smack of the catch, the formalized nonchalance of the heads-down trot in towards the bench, the ritual flips and shrugs and the nervous courtesies of the batter's box. There was a beauty here bigger than the hurtling beauty of basketball, a beauty refined from country pastures, a game of solitariness, of waiting, waiting for the pitcher to complete his gaze towards first base and throw his lightning, a game whose very taste, of spit and dust and grass and sweat and leather and sun, was America.

John Updike (1971), *Rabbit Redux*. Penguin Books, London.

Another way of avoiding the problems of studying sports, but retaining standardized conditions, is to use computer-based games or competitions in the lab, sometimes rigging the results to predetermine who will win or lose. But this raises a similar 'real-life' problem: ensuring that the result matters for the competitors. One way is to offer money, which they win or lose: but the amounts that researchers can afford are usually not enough to make this a life-changing experience. A more subtle method is to make winning a question of prestige, or intelligence, or an explicit ability in some skill or other (e.g. management, financial, or negotiating skills) even though the game may not actually test such skills. The fact that lab games that really have no importance in the lives of the players can be made to seem so is another example of the way the human brain can transfer ancient biological realities—the outcome of a contest in the real world can matter hugely—to an artificial, man-made context without any biological significance, but retaining its social and psychological impact. Those that use these methods often give rather little thought to the skills or experience of their subjects (students are a favourite source); but this can matter. Professional traders playing games that involve money will perform differently from those naïve to financial matters.

Testosterone levels tend to rise in males who win: levels may decline in those who lose, though this is less consistent (Fig. 14). Incidentally, this doesn't seem to occur in women (but see Chapter 9 for a fuller discussion of the role of testosterone in females).

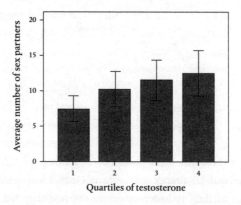

Fig. 14. There is a correlation in later life (>70 years) between testosterone levels and the number of sex partners a man has had. Each quartile represents men divided into four groups based on their increasing testosterone levels.

Testosterone increase is a feature of victory in both sports and computer-based contests; what activates it is strongly dependent on features of a particular society.* Re-living a successful event (e.g. a hockey match) has a similar effect.[100] Testosterone levels are more reactive to winning in some men than others: more dominant and self-confident men tend to have greater testosterone surges after winning. And the identity of his opponent also matters. Increases are more pronounced if a man defeats an 'outsider' (someone he doesn't really know), less if he defeats a friend. So the social context of a victory influences how a man's testosterone will react. This recalls the distinction men make between other males who are members of their group, and those who are 'strangers' (Chapters 5 and 6).

---

* Testosterone may react differently in less sophisticated societies. Chopping down trees to make agricultural plots raised testosterone more than a soccer match in a group of Bolivian tribesmen who live by growing their food in forest clearings. B. C. Trumble et al. (2013), 'Age-independent increases in male salivary testosterone during horticultural activity among Tsimane forager-farmers'. *Evolution and Human Behavior*, vol. 34, pp. 350–7.

But testosterone levels have also been found to increase before a competition, reacting to an incipient challenge. This response to challenge may outrank that to the outcome, and thus complicate interpretation. Men, it seems, raise their testosterone in anticipation of a contest. Testosterone levels in surgeons have been found[101] to increase as the complexity (i.e. challenge) of the operation increases. Furthermore, the winner effect (Fig. 15A & B) can be seen not only in participants, but also in supporters who identify emotionally with the outcome. Thus, testosterone may increase in the (male) fans of winning football teams and, during the 2008 US presidential election, Obama supporters increased or maintained their testosterone levels, whereas they fell in McCain supporters. A study on Dutch males in various corporate workplaces showed that testosterone levels were associated with authoritarian behaviour, but this relationship disappeared once they achieved management positions. So testosterone seemed to be related to striving for promotion, rather than reflecting being a leader.[102]

Do challenge- or victory-induced testosterone surges have any function? There is really very little information, though males tend to behave more aggressively after winning. Whether this is the result of their testosterone surge is not known. The assumption that increased testosterone somehow prepares a man better for a subsequent challenge, or is part of the biological advantage of victory, remains just that. There is some recent evidence that short-term increases in testosterone do enhance a male's liking for risk.[103] We might imagine, if a victory-inspired testosterone surge does increase the chances, if ever so slightly, of a subsequent one, then there may be instances of males being launched onto an upward, testosterone-driven, staircase to success. Defeat might have the opposite result.[†] Since testosterone can act

---

[†] An endocrine interpretation of 'For whosoever hath, to him shall be given, and he shall have more abundance: but whosoever hath not, from him shall be taken away even that he hath'. King James Bible: Matthew 13:12.

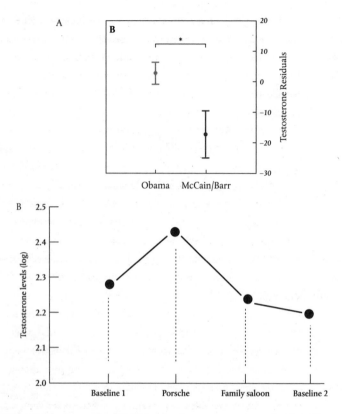

Fig. 15. The winner effect. (A) Differences between testosterone levels in Obama and McCain supporters on the night of the election 2008. (B) Driving a Porsche raises young men's testosterone, whereas driving a family car does not.

quite rapidly on the brain (Chapter 2), it might be that the emotion of 'triumph' which men find rewarding, and which might encourage further competitiveness, could be enhanced by testosterone. This, if it occurred, would be biologically significant, since maintaining the risk appetite of men, particularly young men, makes their social function more effective. This would be another staircase to success. But it is also possible to imagine instances in which heightened testosterone-driven risk-taking, consequent upon a series of successes, might become dis-advantageous, precipitating a dramatic loss.

It is not only defeat that lowers testosterone. Male monkeys that habitually occupy inferior positions in their social group have lower testosterone than more dominant ones; removing the superior males raises testosterone in the lower ranks. Soldiers in the heat of battle have been recorded to have testosterone levels that were not too far above those in castrates; they rapidly reverted to more normal levels once they left the front lines.[104] Stress of any sort is liable to have similar, though maybe lesser, effects on lowering testosterone. This includes being a father,[105] though whether this can be ascribed solely to being 'stressed' could be debated.

At this point, we need to recall the fact that testosterone does not act on its own, and that the physiological reaction to competition is complex. Cortisol, the hormone from the adrenal gland that responds to stress, is important. A challenging situation, particularly if the out-come is uncertain, will tend to increase cortisol. Losing may have the same effect. Cortisol levels increased in McCain supporters (but not in Obama's) on that night in 2008. Both testosterone and cortisol increase in anticipation of a chess match.[106] We need to remember this because the two hormones may interact to affect subsequent behaviour. And there are many other factors to consider, including changes in the autonomic nervous system that controls heart rate etc., and are asso-ciated with the release of yet another hormone, adrenaline (epinephrine).

Studies on how testosterone is implicated in winning or success in real life are quite rare. One needs a competitive, rather homogeneous, situation, in which males strive for such success either against each other, or against a common target. This is difficult to arrange, since lives are complicated and different. But there is one, and it has all the hallmarks we need. It's the trading floor of a large bank, where traders pit their wits, knowledge, and a stream of information to make bets about how bonds or other financial instruments might change, often within seconds. It's a high-risk business; they can make or lose fortunes within minutes. Their income and even their careers depend on their success. There is a considerable failure rate.

Only since 2008 has the phrase 'rogue trader' become familiar to most of us, though it existed well before then. It's a name applied to traders who, by concealing their losses from their bosses, or using other illegal methods of trading, perpetrate the classic gambler's strategy: chasing loss. Having lost a considerable sum, they then attempt to regain it by a bigger (and, usually, riskier) gamble, which nearly always results in an even bigger loss. Hence one rogue trader who bankrupted Barings, a bank that had existed since 1762, and another who jeopardized the French bank Societe Generale; he lost even more, but the bank survived. Most financial traders are not rogues; but they are all gamblers.

Rather few people have ever seen a trading floor (they don't like strangers around), but if you were to accompany me on a visit, several things would strike you (Fig. 16A & B). There is a huge room, filled with rows of traders. Each trader sits in front of several (six or more) computer screens and has a telephone. The screens change constantly, and each shows something different. In the background, a loudspeaker relays economic news and information. The place is rather quiet, though every now and then a murmur goes round the room (if you listen carefully, it consists mostly of expletives). Most traders don't seem to be doing very much, though every now and then they type furiously for a minute or so. They watch all their screens constantly. Look a little closer, and you'll notice something else. Nearly all the traders are young (or youngish) men. There are very few over 40. There may be one or two women, but you'll need to search the room to find them. Before the electronic age, trading floors were much noisier, more obviously emotional affairs. Traders would stand around, shouting prices or deals; the energy and noise levels were intense. One or two are left, but nowadays mostly it's a quieter, computerized environment. Nevertheless the underlying tension remains.

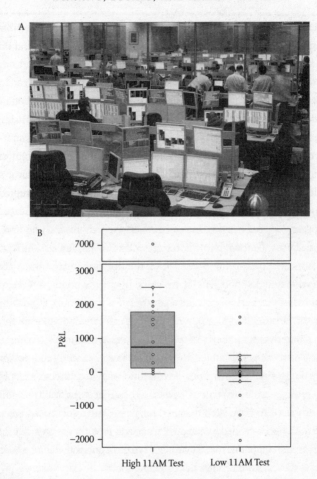

Fig. 16. (A) View of a financial trading floor. Note the crowded environment, and the number of screens each trader has to monitor. (B) Traders on days when their morning testosterone was higher made more money (P&L: profit and loss) than when their testosterone was lower. The dark line is the median value of their P&L. The dots are values from individual traders. The shaded boxes are a statistical measure of variation (confidence limits).

There's a very simple reason why most financial traders are young(ish) men. The nature of trading incorporates all the features for which young males are biologically adapted. We have seen how young males are attracted to risk, and why this is so: they need to take risks to gain mates and the other accoutrements that accompany reproductive success. For the same reasons, they need to be competitive, and take rapid decisions on the basis of whatever information they have, both current and gleaned from previous experience. Traders spend their time taking rapid and fateful decisions. Whether they trade in bonds, commodities or shares, or anything else, they are trying to predict the future. They are making an estimate of future value; if they get it right, they win: otherwise they lose. If you know the value of something in the future, and it's more (or less, in some circumstances) than the present one, then buying it now will make you money when you sell. That may be in a few seconds, a week, or months or years ahead. Traders characteristically 'close' their 'position' (i.e. sell) at the end of each day, though this can vary. Not only are they are taking risks with their money (or the banks); they are also competing against each other, since the managers keep a careful score of how they do, and persistently poor performance may lose them their jobs. The rewards can be enormous, but so, too, can the penalties. The whole set-up seems to have been designed for young men. All the actions of testosterone are echoed by the qualities required of a successful trader. It does seem remarkable that the artificial world of financial trading should so suit the innate characteristics of young males.[107]

But not all young men would make successful financial traders. It's not altogether clear what predicts success, but formal education is less important than one might think. On a London trading floor you would find those with PhDs in mathematics, but also those who left school at 16 and worked in market stalls (the others call them 'geezers'). I have no idea how common the necessary qualities are, but I'd guess that if you stopped 1,000 young men in the street, you might be lucky to find one who could survive the rather brutal world of the trading floor. Their testosterone levels would not be a reliable guide. So it's not just youth and testosterone, but some other quality as well. With the recent advances in genetic technology, whereby an individual's DNA can be profiled quickly and quite cheaply, it would be fascinating to see whether certain genetic variations predict those who might do well on the trading floor.[108] For this is Darwinism at its most dramatic. The fittest really do survive, and the less fit go under. If there were to be genetic variants that did predict success then the question would arise: does this profile (or profiles—there are likely to be more than one) also predict success in another, more natural, environment, long-standing enough to have acted as a positive selection process? Is there a parallel in the natural world for success on the trading floor? Is the trading floor the modern equivalent of the jungle? The shape and nature of the traders' world reflects the properties of the male brain, unlike the natural world, which owes nothing to the male brain, but played a major role in determining its abilities. So there might be rather special genetic predispositions for success as a trader.

There are some mutations that might qualify. One is in a gene that controls serotonin. It alters the rate at which serotonin is removed from the synapse, and one form of this gene is associated with (among other things) increased 'impulsivity'—the ability (or lack of it) to restrain a response to a particular stimulus. A common way of testing this is to offer subjects the choice between two (monetary) rewards: a smaller one immediately, or a larger one after a delay. Both the relative amounts and the delay are varied. More impulsive subjects choose the immediate reward after being offered shorter delays or smaller delayed amounts than less impulsive subjects. That is, they can't bear to wait. Testosterone has been associated with increased impulsivity, and may interact with serotonin genes. Another is a gene that regulates how the brain responds to dopamine: a particular variant increases what is termed 'sensation-seeking'—the liking for stimulating and exciting experiences (Chapter 10 has more information on this). Testosterone interacts with both these genes, and many others. This is likely to have relevance to financial trading as well as many other instances of risk-taking. It would be a mistake to think that testosterone alone shapes this behaviour.

One can't help comparing the risky behaviour of young traders to those in ancient times, creeping through the trees, wary of enemies, gathering what information they could, devising risk-laden strategies to kill or capture their enemies, with riches in the form of food, territory or females as the prize for success, but failure promising starvation, injury, or death.

One of the surprising facts about young men who employ traders is how little they know, or seem to care, about what goes on in the brains of their employees. Investment bankers know a great deal about the markets, and the statistics that can influence them; they know about the technicalities of swaps, and the mathematical models that underlie much of present-day finance. But they seem curiously incurious about how to predict a good trader at any level other than simply trying him out in various ways. The range of tests used, including those on personality, certainly give prospective employers some idea of their applicants' ability, but these are at the levels of function. The underlying biological or physiological properties that might contribute to these qualities remain unexplored. The great revolutions in biological, psychological, neurological, and genetic knowledge have, it seems, passed them by. One day, maybe, they will begin to be curious about the biological make-up of the deci-sion-making human computer that sits in the chair in front of the electronic one. The two work in quite different ways.

Though testosterone may predispose males to be competitive, or take risks, does it play any part in the actual process of making financial decisions? If such decisions were all made rationally and in the light of sufficient information, entirely to maximize gain, it would be difficult to see a role for testosterone. Economists have hypothe-sized such a person: the rational man. Theories based on such an individual have signally failed to predict actual behaviour. He doesn't seem to exist in the real world, only in the minds of economists. More recently, economists have realized that decisions are often based on incomplete, or misinterpreted information, or biased points of view, or emotional, rather than rational, attitudes. A simple example: if you ask people to bet on an outcome that has a 50% chance of winning, they do so more often than one that has a 50% chance of losing. The two probabilities, of course, are the same. The difference is the use of the word 'winning' or 'losing': most people are what economists call loss-averse: they dislike losing more than they like winning. So the words 'win' or 'lose' bias their decision.[‡] Another test is to ask subjects to choose between two outcomes: one has a high probability of yielding a low reward, the other gives a greater reward but occurs less often. By varying the balance between the two, one can determine how the subject balances the amount of reward with the likelihood of success.[§] Males consistently prefer riskier options than females (i.e. their balance point is more towards higher reward, lower probability). Recent evidence (using functional magnetic resonance imaging (fMRI)) suggests that there is a balance between two systems in the brain during this kind of decision-making: one concerned with emotional responses (focused on the amygdala) and another (the 'analytical' part) on the area of the cortex called the anterior cingulate gyrus (it lies just behind the frontal lobes).[109] This (like the amygdala) is part of the limbic system, which is particularly concerned with survival (Chapter 10 describes the limbic system in more detail). We are begin-ning to understand what happens in the brain to account for the fact

[‡] It's called a 'framing' effect. Numerous other examples, and a wider discussion of these topics, are given in D. Kahneman (2009), *Thinking, Fast and Slow*. Allen Lane, New York.
[§] It's called 'prospect theory'.

that so many of our decisions, which we like to think of as entirely rational, are actually heavily coloured by emotion. For example, economists have long known that 'value' (economists call it 'utility') of an outcome is a powerful incentive and guide to decision-making. The value of a given object (say, a sum of money) varies with the attributes of the recipient (a poor man will value £10 much more than a rich one). Recent evidence suggests that those parts of the brain known to be associated with emotion (as well as reward) are particularly concerned with assessing value, and the way that this may change for an individual under different circumstances.[110]

Testosterone does seem to make risk-taking more attractive, and this has obvious relevance to the world of finance. There is a large literature on the psychology of taking risks, and the way individuals evaluate risk in relation to possible benefits.[111] There seem to be two mental processes, variously termed 'emotional' vs. 'logical', 'experiential' vs. 'analytical' or 'fast' vs. 'slow', etc. The first is based on the emotional (or 'affective') response to a particular situation, based—at least partly—on previous experience, which associates it with a particular emotional state: this requires little analytical or reasoning power, so occurs rather quickly. If a reliable decision can be made in this way, without the need to analyse the situation slowly, this has evident benefits, particularly in cases of urgent need. A classical example is the way someone will avoid stepping on a snake. But previous experience may be an unreliable guide, as may be the emotion associated with it, so some circumstances are better dealt with by a slower, more logical appraisal of what has to be done and the risks associated with it. This depends both on the reliability of the information currently available and the person's ability to draw rational decisions from it. In reality, it seems that most decisions are a mix of the two processes,[112] though many men may underestimate the influence of the first, more emotional, response. I do not know of any firm data showing which component of decision-making is regulated by testosterone: but it's an obvious guess that it will be the first, 'fast', emotion-based process that will be most sensitive to its effects, particularly, we may suppose, in younger men (but see later in this chapter for another interpretation). A liking for risk is one example; aversion to loss is another. Both are testosterone-sensitive. It took psychology to show economics that 'rational man' (a 'slow' thinker) was an incomplete and, indeed, unreliable model for financial—and other—decision-making in the real world.[113] Now the new science of 'neuroeconomics' is attempting to bring what we know about the brain to enlighten our understanding of how financial decisions are actually made, and how different brains vary in their ability for doing so.[114]

If testosterone does encourage the emotional, 'fast', 'experiential' basis of decision-making, when might this be an advantage? One answer is: in precisely the situations in which young males are likely to find themselves. Those in which emotions run high, in which rapid decisions are required (e.g. in a fight), and which have formed a consistent feature of the young man's life (so there is plenty of precedent). Competing for mates or assets (an individual activity) or defending the group (a corporate one) are substantial ingredients of life of the young males of most societies: surely it seems consistent with what we know of the function of testosterone to suggest that biasing young males in the direction of this type of decision-making might well be a biological and social advantage. Exactly what advantage will depend on the cost/benefit ratio: how many losses can be sustained in order for there to be a reasonable probability of a major success. Transfer these ideas to the modern world, and the trading floor is an obvious example.

A study on a London trading floor showed that male traders made more money on days on which their morning testosterone was higher than usual.[115] As scientists never tire of pointing out, a correlation is not necessarily evidence of causation, though it may be. But the first question is: why is testosterone higher on some days than others? In this study, it was not due to their having made more money than usual on the day before. Was the challenge of a particular day greater than normal? We do not know. Neither do we know whether testosterone itself was influencing their decisions—for example, by altering their appetite for risk ever so slightly. There will be an optimum strategy which may change from day to day: too much risk will result in larger losses, too little in not enough gains. Though it is generally agreed that males are less risk-averse than females, giving subjects testosterone has had mixed effects on financial risk-taking, insofar as this is measured by playing computer games that mimic, in various ways, actual financial transactions. Subjects in these studies, unlike traders, were not professionals. It is important to recognize that the processes of financial decision-making may vary with both the context of the decision (rapid, considered), its significance and who makes it (a professional, practised trader, or Mr and Mrs Smith buying a car every 3 years or so, or Mrs Smith shopping at her local supermarket). Subjects (usually students) playing a financial game in a lab are different again. Economists often try (rather implausibly) to devise theories that embrace all these situations and more. And there are always questions about the dose and duration of testosterone treatment, and the sensitivity of the tests used. Relating risk appetite to individual variations in testosterone levels has also given inconsistent results. The 2D:4D ratio—a controversial index of individual exposure to testosterone in the womb (see Chapter 3)—has not been much better as a predictor of risk appetite, though there are reports that both men and women with lower ratios (i.e. more male-like) prefer riskier behaviour.[116] But there is the interesting suggestion that a low digit ratio (more early testosterone) is associated more with heightened reasoning ability (the 'slow' mechanism) and it is this that may be responsible for greater appetite for risk.[117] The more someone can analyse a situation, the more he will be able to assess the actual risk. The rational assessment of optimal risk may improve a trader's chances of survival. No-one has measured genetic variations in the testosterone receptor in traders and how it relates to success, though this may be crucial (Chapter 2).

Some individuals are more adaptable than others, and some perform better in one set of circumstances than another. The more uncertain an outcome, the greater the risk. Cortisol, the 'stress' hormone from the adrenal gland, is particularly likely to increase when someone is required to deal with a difficult or threatening situation, but one in which there is considerable uncertainty about the nature of the demand or the way to deal with it. Markets vary in uncertainty and variability (there is a measure for this). The traders we studied showed increased cortisol on days when the market was highly variable and therefore increasingly unpredictable. This occurred independently of changes in testosterone. Cortisol has many actions, but one is to enhance anxiety, and higher levels tend to strengthen the impact (memory) of a fearful event or its possible consequences. Giving cortisol to young men (students), playing games that mimic trading floors increased their appetite for risk. Testosterone, on the other hand, also increased risk-taking, but this was because they felt more optimistic about the results of their actions.[118] Cortisol and testosterone interact, so if we are to form a clear picture of ways in which a trader's hormones influence his decisions, we need to know about both. But there may also be genetic patterns that make a trader better at dealing with one situation than another (e.g. a 'bull' or a 'bear' market). We (and their managers) know nothing about such matters.

The focus on traders must not blind us to the fact that there are many other jobs in the financial world, many of them requiring critical decisions, but under rather different conditions or circumstances than those of traders. Many traders do not make good managers, which suggests that the qualities that make a good trader may not necessarily be those for other financial occupations. Does testosterone play any role in the other players in the financial world? It is a world dominated by men, but we have no information. And then there are the rest of us, who make periodic decisions, financial and otherwise: do the same rules apply to us, and are they the same under all circumstances? Does testosterone, either in the womb or in later life, play a significant part? What we know of the behavioural effects of testosterone suggest that it might well do so. There's a lot more work to be done!

Generosity, trust and a sense of fairness are all part of many financial transactions. The 'ultimatum game' tests this. Subject A is given, say, £10. He is told that he can offer any fraction of this to subject B. If B accepts, then they both take their share. But if B refuses, then neither is paid anything. Subjects usually accept offers above about 30%, but giving testosterone to subject A makes him less generous, whereas treating subject B makes him less likely to accept a lower offer. In other words, they both become meaner.[119] Giving testosterone to women had no effect on their ability to do a visual task, but disrupted efforts to make them collaborate: in other words, they become more ego-centric.[120] Such studies give us further clues about what testosterone may do in the financial world as well as in the rest of life.

Interesting and informative as these laboratory studies undoubtedly are, we lack enough data on real-life situations. They may be difficult to arrange, and the multiplicity and complexity of the financial world and those who work in it may mean that results obtained under one condition may not necessarily apply to another. Nevertheless, there is no substitute for real-life observations, and technical advances in the way that data can be obtained without interfering with what people are doing are making this objective all the more realizable. Another problem, as this chapter shows, is that much of the information we have, or could get, is an association: levels of testosterone, or the variant of a particular gene, or an fMRI picture may be associated with performance, ability or some other measure: the problem is always to translate this into a causative mechanism. The customary method of doing this is to change whatever one postulates is the underlying 'cause'—for example, levels of testosterone—and confirm that what one measures has changed in the predicted way. In some situations involving humans in real-life situations (e.g. financial trading) this is either difficult or impossible, for ethical, legal, practical, or technical reasons. There are ways of getting round some of these obstacles: for example, getting professional traders to make financial decisions as part of a 'game' but which they are made to think is important for their career, and relating results to their real-life performance. The banking industry (and many others) badly needs to know more about the people it employs, and on whose performance its future (and ours) rests.

Testosterone also operates on qualities such as competitiveness or aggression, and these have consequences for how decisions are made. We will consider how testosterone acts on the brain more closely in Chapter 10, but here we can note that the frontal lobes of the brain are concerned not only with emotion, but also with planning, social awareness and introspection. Damaging them makes men indifferent or unaware of risk and, both in the lab and in real life, they make disastrous economic decisions. It is likely that most of the functions of the frontal lobes are independent of testosterone, but moderate, fine-tune and direct actions that testosterone can also influence by its action on other parts of the brain. Interactions between those parts of the brain directly sensitive to testosterone and others may not be limited to the frontal lobes. Since decisions are so varied, both according to their context, their nature and by whom they are made, very wide areas of the brain will be recruited. These include other parts of the cortex** and the memory stores, which are still poorly

** The parietal lobes, which lie behind the frontal lobes, have been repeatedly implicated in decision-making. So has the cingulate gyrus, part of the cortex closely associated with the limbic system. Chapter 10 deals with this is more detail. There is an expanding literature on the parts of the brain involved in financial decision-making, based mainly on imaging (fMRI). Many of these areas (e.g. the frontal lobes, amygdala, etc.) or systems (e.g. dopamine) are implicated in other functions as well (e.g. planning, reward, mood). For some recent reviews see *Current Opinions in Neurobiology* (2012), vol. 22. How hormones fit into financial decision-making is not discussed.

understood, but may involve wide areas of the brain, essential if decisions are to take experience as well as current circumstances into account ('slow' thinking). Also included are parts of the brain known to be associated with reward, essential for the control of eating, drinking and sex, but in humans able to derive rewarding stimuli from many other activities, including getting richer. And, as we have already noted, the human brain has manufactured the financial world, though (maybe unwittingly) in a form that is full of the resonance of a more ancient world and its needs.

The great financial crash of 2008 inevitably resulted in a chorus of claims that, had banks been run by women, this would not have happened, and that future disasters could be eliminated by recruiting more females into executive positions. There should be no obstacles in the way of women who have the desire and ability to reach such positions. But this facile remedy ignores the root of the problem: it was not excessive testosterone (or too many males) that led to the banks failing, but lack of control over behaviour which, indeed, may well be influenced by testosterone in many ways. But as Chapter 6 amply shows, without control on its action, which is universal, un-bridled behaviour of any sort, including that sensitive to testosterone, is incompatible with a stable and functional society. And so it proved in the financial world as in any other.

It does seem extraordinary that the modern world of finance, enrobed in steel and glass, replete with computers, electronic aids of all sorts, artificial systems of commerce (i.e. money[††]), elaborate mathematical theories, complex managerial structures, so remote from caves and forests, is still redolent with a more ancient force, carried along the centuries. For this world, like every other in which men strive to succeed and survive, is permeated with the manifold

[††] A classic discussion of the function and sociology of money is give by Georg Simmel (1907), *The Philosophy of Money*.

actions of testosterone. There is no chance of freeing finance from testosterone, or all the other biological, psychological and neurological influences about which we still know so little. But greater understanding would give us greater control. Financial decisions, since they represent potential gain, are very similar to monkeys competing for a fruit tree. The difference is in the complexity of the context in which they are done, and the evolved neural mechanisms for assessing and estimating decisions in the more elaborate environment of the artificial financial world. Testosterone has the same function in both situations, by encouraging males to be competitive risk-takers.

The powerful impact of testosterone is not only on individual men, different as they are. It is also evident on the way that groups of men behave together, and how they interact with other groups. In some circumstances, this can lead to war.

8

Testosterone and War

In 1978 the anthropologist Carol Ember calculated that 90 percent of hunter-gatherer societies are known to engage in warfare, and 64 percent wage war at least once every two years...In 1972 another anthropologist, W. T. Divale, investigated 99 groups of hunter-gatherers from 37 cultures, and found that 68 were at war at the time, 20 had been at war five to twenty-five years before, and all the others reported warfare in the more distant past. Based on these and other ethnographic surveys, Donald Brown includes conflict, rape, revenge, jealousy, dominance, and male coalitional violence as human universals.

> Steven Pinker (2002), *The Blank Slate*. Allen Lane, London.

The soldiers sprang to their feet and charged, and simultaneously the second machine gun opened fire....In the incomprehensible hurricane of bullets the soldiers whirled and fell for half an hour ...Coldly Hectoro dismounted and walked among the carnage, slicing the throats of all who still lived...Back on the field of slaughter the victors were both jubilant and appalled. Shaken, pale and trembling, they embraced each other and then wandered dumbly among the fallen.

'They were innocents,' said Misael. 'Look at them, they were all boys.'

'Yes,' said Pedro. 'Little boys with mad leaders and fear in their hearts.' Louis de Bernieres (1990), *The War of Don Emmanuel's Nether Parts*. Martin Secker and Warburg, London.

This is not a book about war, or the causes of war. There are plenty of those.[121] War has existed since the earliest records of mankind, and no doubt well before then. Philosophers have long debated how to define 'just' or 'unjust' causes of war (*jus ad bellum*) and whether there are ethically defensible ways of pursuing them (*jus in bello*). These are ethical questions of great importance, but outside the scope of this book. An extraordinary feature of those reams of words written by philosophers, as well as by journalists, historians, and politicians about war and its dreadful nature is the almost complete lack of interest in the biological make-up of those who wage it and whether this contributes to the phenomenon of war. Discussions on the causes of war (and whether these are justifiable) are often based on terms of the objectives of war—claims on territory, or fear of incipient attack, or the preservation of democracy or humanitarian aid for instance. There is another level: what is it about the biological constitution of mankind (or other species) that inclines him to war, and what are the circumstances that promote war as a biological strategy to improve 'fitness'? This may be one consequence of the traditional separation of disciplines, and is an increasing peril as disciplines become ever more complex. For example, most philosophers who write about war have little interest in biology or neuroscience.[122] There are even those who deny that detailed understanding of the brain is useful for shedding light on the mind (a traditional province of philosophy). Biologists and psychologists have not ignored war, but approach it from their points of view. Philosophers would no doubt point out that biologists need to know rather more philosophy. Just as economists and bankers seem uninterested in the constitution of those who operate as financial traders, or determine a bank's policy, or make national economic decisions, so analysts of war have, for the most part—but not entirely—ignored the biological and neurological features of mankind that predispose him to war. In earlier times, war was indeed glorified, in the face of the dreadful consequences it often had both for individuals and their societies. More recently there has been a more universal acknowledgement of the frightful nature of war, though little sign that this has reduced its likelihood. One pillar supporting the formation of the European Union was the desire to prevent the European-derived wars that so disfigured the nineteenth and twentieth centuries. The Second World War (1939–1945) was the bloodiest in history (60–80 million deaths). By contrast, it is said that 3 million died in the Hundred Years' War (1337–1453). How far does what we know of the biological underpinnings of mankind account for these attitudes to war?

No-one can write anything, it seems, about war without referring to Carl von Clausewitz's famous book *On War*, a work that is still required reading for soldiers, politicians, philosophers, and anyone else interested in enquiring about the reasons for which wars occur, and what influence they have had on human history. In this work, he proposed his famous 'trinity':

> War is a fascinating trinity—composed of primordial violence, hatred, and enmity, which are to be regarded as a blind natural force; the play of chance and probability, within which the creative spirit is free to roam; and its element of subordination, as an instrument of policy, which makes it subject to pure reason.[123]

This has been interpreted in many ways, but note that there are several phrases in Clausewitz's idea which are of the greatest relevance to a book such as this: 'primordial violence . . . blind natural force'—which infers that there is something biologically basic about violence as a behaviour and hence the propensity of war; the concept of subordination as an 'instrument of policy' (i.e. as a way of influencing social structure or relationships between groups or nations); and the way that using this demonstrates the role of something he calls 'pure reason', implying that this exists in contrast to emotional reactions and 'primordial violence'. As we have already seen, it's not so easy to separate emotion from reason, and to allocate them distinct roles in decision-making.

The social structure of those species of monkeys and apes that live in multi-male groups represent, as already described in an earlier chapter, tense organizations in which individual status is determined by gender, previous outcomes of aggressive encounters, liaisons, sexual readiness, and so on. Between the males there exists an uneasy peace. We have seen that access to food, mates, and advantageous parts of the habitat is determined largely by a male's established status, without recourse to repeated testing of the outcome of a dispute. Any attempt to disturb the status quo will result in a threat from a more dominant male, or even a fight, if the threat does not suffice. But the status quo is not a stable one, that is, accepted without testing by each male. So any weakness on the part of a dominant male is rapidly detected, and may result in rearrangement of the social hierarchy: peace and the group's social structure are always provisional. Life is not tranquil for anyone, particularly the adult males.

Now suppose a neighbouring troupe appears, and attempts to feed in the trees that are normally the province of the home troupe, and thus invade its territory. The reaction of the males of that troupe is immediate and striking. All combine to repulse the newcomers. Whether a male is a large, dominant one with formidable canine teeth, or a smaller one not so long out of puberty, they advance together to repel the invader. Individual status is seemingly forgotten: all of them act as if there is a common threat, irrespective of their position in the group. No longer does a smaller, subordinate male fear to go too close to a larger dominant one—all focus on the job in hand: to secure the group's resources.[124]

This scenario has a number of features. First, the males of a group recognize the other members of their group, and treat them differently from the invading males: you could say they are exhibiting group allegiance. Second, it is the males who defend the group's territory or resources. Third, though all the males are engaged in activity—fighting—that involves considerable risk, they are not doing it for immediate personal gain, rather for the welfare of the group (though, of course, this might hold eventual benefits for them as individuals). It is by no means certain that, at the end of the encounter (a primate war, no less), every male will benefit in any way. Thirdly, it is males who are intolerant of the presence of other males in a particular area (some biologists would call this 'territoriality', but there have been disputes over the exact meaning of that term); it is debatable whether this involves a cognitive realization that yielding territory would also lose the group some of its assets. All that is biologically necessary is that the males of one group recognize some aspect of their habitat as 'theirs', and have inbuilt intolerance to sharing it with other monkeys (apes) who are not members of their group. In other words, they have group identity and behave differently towards their fellow members than to strangers. This represents a kind of bonding and they act together in mutual support in defence of their assets, or in attempts to gain assets from other groups, despite the attendant risks and their competitive attitude towards other members of their group during more tranquil times.

These attributes have important biological consequences: they ensure both that the resources of their habitat remain accessible to the group as a whole, and also that the females of the group will continue to be available to be fertilized by the group's males. If the group is defeated, it may lose both. There is a separation, therefore, between the immediate causes that make male primates go to war (territorial intolerance), and the more fundamental biological consequence of those reasons (access to essential resources). It is the latter that acts to ensure the biological 'fitness' of the group. The idea that groups could show collective 'fitness' and were thus one basis of selection—group selection—was proposed particularly by V. Wynne-Edwards.[125] It attracted considerable criticism, and was largely discounted for many years, and there are still biologists who would deny its validity. Now, however, it is an idea making a considerable comeback. It is important, therefore, to distinguish between those factors that predispose towards aggression between members of a group from aggression by a group as a whole towards another group. It is the latter that qualifies as war. Clausewitz has been criticized for limiting 'war' to conflict between nation-states; but the essential elements are apparent in any organized conflict between two established groups—that is, two groups whose members relate to each other in the special way (group identity) shown not only by non-human primates, but by men (particularly men) as well. Aggression within or between groups is often confused, simply because both seem, at first sight, similar: they involve fighting, and both are conducted for some sort of gain or defence.[126] But the argument here is that they are different neurobiologically in important ways, though they may both have, as one of their determinants, a common factor: males are very likely to do both. Can we substitute the word 'testosterone' for 'male' in this context?

Chimpanzees live in groups that regularly interact aggressively with other groups, to protect or obtain territory, food or females. Such forays are made by groups of adult males, not females. The average body weight of a male is around 40–60 kg whereas for females it is 30–50 kg, so that the extreme gender differences in body weight seen in some other primates does not apply to chimpanzees, and cannot, by itself, account for the marked gender difference in the participants in group aggression (though it may be that females lack some of the males' muscular power as well as their teeth). An appreciable number of male chimpanzees are killed or seriously wounded in boundary fights. Chimpanzee war has been offered as a model for human hunter-gatherer war,[127] and thus, by implication, for the evolution of modern warfare. How far is the way males wage war dependent on the role of testosterone?

Psychologists have added another dimension. They have asked: what is it about human psychology that enables war to occur, why is it so frequent, and what are the circumstances in which it is likely to occur? Edward Tolman,[128] known for his pioneering work on learning, recognized the difference between individual and group aggression, but showed how an individual's case (e.g. a rival group stealing his wife[†]) may result in the aggrieved individual soliciting support from other members of the group and thus instigating a war, as in the Homeric account of the Trojan War. This example is highly relevant to the subject of this book, involving, as it does, cooperative action by the males in a group and behaviour that has reproductive implications. Testosterone is central to the latter and it is characteristic of young males to form cooperative bands (for example, street-gangs—see later in the chapter) in response to a common threat. This seems to be another property of testosterone, essential for group survival.

There is an uncertain boundary between a situation in which a society identifies a subgroup within it as an enemy (e.g. anti-semitism) and how it views an outside group. In fact, the difference may be less than it appears: by demonizing an identified set of people, a society effectively denies them membership of that society. So they are treated as 'outsiders' and the definition of 'war' applies. But there is general recognition that personal pugnacity is not sufficient to account for war. Indeed, soldiers in battle may show little hate or aggressiveness towards the enemy,[‡] but are more likely to experience fear, homesickness, or even boredom. They don't fight for their country so much as for each other: their colleagues in the combat group (platoon, etc.).[129] Wars are initiated by leaders, who are nearly always (older)

[†] For example, Helen of Troy.
[‡] A famous example occurred in 1914 during the First World War, when there was an unofficial truce at Christmas between German and English soldiers. For four days they talked, exchanged gifts and addresses, and only resumed fighting on the orders of their senior officers.

men.[130] Armies are directed by generals, also usually older men, but those who actually do the fighting are, mostly, young men. About 75% of those killed in action during the Second World War were under 35 years old. The recently increased role of (Western) women in actual warfare does not alter this historical fact. Despite their common masculinity, the motivation for collective, organized fighting (war) in leaders, generals (the two may be the same in some circumstances), and soldiers is different. But all are influenced in some way by the actions of testosterone, as we will see.

Scrutinizing the phenomenon of warfare in primitive (by which is meant technologically and culturally less sophisticated) societies, such as hunter-gatherers, allows us to explore the role of human males more closely, relatively uncontaminated by the complexities of modern society and technology. The hope here is better information about the roots of war, though we should always be wary of confusing the evolution of war (a generation-dependent event) with cross-cultural comparisons (these assume that present-day primitive societies are similar to those that existed more generally in the distant past). This is similar to the fallacy of confusing comparisons between species with the evolution of one species (see Chapter 1).

Studying such societies across widely dispersed areas of the world (Australia, Africa, South America, etc.) shows some constant features. The first is that warriors are prized, and fighting prowess is prized especially. This gives high social status to such individuals, with consequent positive standings in political, social, and sexual contexts. In earlier times, kings were expected literally to lead their armies into battle. Only later did it appear more prudent (and effective) if they, and their generals, directed attacks from behind the lines. But often they were still painted at the head of their troops, brandishing a sword. There is thus strong cultural encouragement for young men to develop such skills from an early age. Regular hunting and ritual combats provide practice and competition for young men who would become military leaders. Has there been natural selection, not only for warlike physiques, but also the liking for war or for taking risks? There does seem to be an appetite in young men for the excitement of war, noted by many observers of more complex societies as well as less-developed ones.[131] Most wars among these tribes involve attempts to win or defend territory, or capture future wives, similar to the situation in chimpanzees. But wars can be fought for revenge, something that is difficult to study in chimpanzees. It seems that, in many cases, there is little personal gain (other than prestige; though pillage is an exception) for the males who take part in war. One group, the Yanomami of South America, has been dubbed 'the fierce people', a prime example of 'primitive' conflict and a presumed example of Hobbes' 'state of nature' (the result of an anarchic society—see Chapter 6), but this is disputed.[132]

Move along time to more complex societies, and other features of war appear. Uniforms proclaim a soldier's group (his regiment) and his status in that group (rank).[133] An army is the most obvious example of formalized social rank. There is no room to dispute or challenge this rank: it is settled by law and reflected by display. Only very occasionally, and then under extreme conditions, do soldiers kill their own officers. Up to the nineteenth century, generals were often appointed because of their social (civilian) rank: in England, dukes most often led the battle. This had periodically disastrous results: so now leadership has been professionalized, a good example of how human social and intellectual development has overridden a more primitive social order. But there were exceptions, such as the Duke of Wellington, a renowned general who defeated Napoleon at Waterloo. Bernard Cornwell's fascinating account contains vivid descriptions of the savagery and brutality of battle.§ But this has

---

§ Bernard Cornwell (2014), *Waterloo: The History of Four Days, Three Armies and Three Battles*. HarperCollins.

had other effects: while in some societies it has led to those making decisions about war being separated from those who conduct it, in others it has given those who command armies social and political power as well.

Uniforms do more than signal regiment and rank: until rather recently, they were resplendent, their vivid colours reminding one of the displays of males of many other species during the testosterone-controlled breeding season. Their design promulgated maleness and masculine status: epaulettes broadened shoulders; helmets, shakoes, and bearskins accentuated height—all attributes of the testosterone-formed body. Only after it became clear that the invention of guns had made colourful uniforms a distinct disadvantage did soldiers wear khaki and helmets for protection rather than display. But they still revert to colourful dress on peaceful regimental occasions. Military uniforms, by all accounts, still confer on their wearers considerable status and sexual attraction, recalling the social value of military prowess in hunter-gatherers.[134]

There are striking parallels, therefore, between human war and organized fights in animals, particularly non-human primates. Much has been made of the supposed difference in lethal combat between chimpanzees (and other animals) and humans. Only humans, it is said, have the propensity to kill each other, and here lies an important biological distinction. But is this true? Careful observation of chimpanzees has shown that they, too, will kill rival males.[135] The real difference lies in weapons. Chimps have only hands and teeth (though they have been seen to use rocks and branches as weapons): it is comparatively easy for a male chimpanzee to run away from an aggressor and be safe. But for a man to run away from another armed with a spear, a more lethal bow and arrow, or an even more deadly gun, does not guarantee safety. Chimps are simply not very good at killing: men are much, much better. The difference between the two species is more technology than motivation.[136]

War occurs not only between nations. It is everywhere on the streets of our larger cities. And it has much the same features as other types of war. The street gangs of Los Angeles have been much studied, and it is said that 94% of US cities with more than 100,000 inhabitants have such gangs.[137] Characteristically they are gangs of young men, usually of the same ethnic or racial origin, and from a comparatively deprived social and financial background. Entry is difficult, though often highly sought; membership brings added self-esteem, but also access to money and other assets, including women. Potential members may spend time as fringe members before being admitted. Gangs defend territory, and are particularly hostile to others of the same origin (since they represent the greatest threat). Individual status, or reputation, is highly prized, and is based largely on prowess in fighting. Many street gangs, but not all, are involved in criminal activity. In 2006, over half the reported homicides in Los Angeles were attributed to gangs, many the result of inter-gang warfare. Females are seldom members of such gangs but may be 'hangers-on'. All-female gangs are very rare. Gang members may wear special clothes, or symbols, or use words or signs peculiar to the gang. Assessing the number of gangs is notoriously difficult: in 2006, London was said to have around 120 gangs and New York about the same number (both are almost certainly underestimates).

For many deprived young men, gangs offer excitement, group identity, self-esteem, and social status for those that lack them, as well as a common purpose: defending the gang's territory, for example, but also, in some cases, material and sexual gain. But the propensity of young men (and, indeed, older ones) to form 'gangs' is not limited to the streets of deprived neighbourhoods. A football team can take on many of the features of a street gang. In both, entry is limited and subject to defined requirements; other teams may be viewed as 'enemies'. The role of territory is obvious: football teams win more often at 'home' than 'away'. 'Bikers' form gangs focused on their machines, and may be hostile to other groups with different ones. At the upper end of the social scale, men form clubs (though, in this case, not limited to the young) that restrict entry, and award privileges (e.g. social status) to members, and may also have a common objective (e.g. science, literature, politics, etc.).[138]

There has been a tendency to regard street gangs as a distinct social phenomenon, characterized by working-class or deprived socially-deviant criminal activity. In fact, though there are certainly special features of street gangs, much of their structure, organisation, and purpose reflect a more general propensity of males—particularly younger males—not only of other classes, but also of other species. The observation that the males of social groups of certain non-human primates act in concert to defend the troop against invaders, or to attempt to seize the assets of other groups, has already been mentioned. This is not limited to primates. Similar behaviour in humans has been recorded repeatedly in, for example, the inhabitants of particular villages, tribes, or other communities. Members of professions, companies, institutes, and other work-place or social organisations (e.g. cliques, golf clubs) are all members of gangs, in the sense that these provide an identifiable social structure, an inherent hierarchy, intra-group rivalry—but inter-group competition—and a sense of 'belonging'. They also, in many cases, provide avenues to livelihoods. Access to a particular 'gang' is most often limited, and disloyalty, rule-breaking, or under-performance is punished in some way. Until recently, most of these 'gangs' have been male-dominated or even exclusively male. Street gangs (which have many forms) are only different in the sense that they use non-conformist methods to achieve similar objectives. Thus, while they have admission procedures, rules, and hierarchies, as do more conventional 'gangs', their objectives (most recently, largely drug-related) are not those accepted by a wider society, and their methods of control (often extreme violence) are also beyond what the rest of society deems reasonable. They do not accept the legal norms of that society, but they have their own rules nevertheless; they are not anarchic. Much effort by sociologists, criminologists, and others has been expended on understanding why street gangs occur (they are not, as is sometimes thought, an exclusively American phenomenon), but a central fact is that this is only one expression of a more general biological property of males, essential for effective social function, but only aberrant according to the norms and rules of the current society in which they occur. These norms have varied between cultures and historical periods in humans (a function of the adaptive and inventive ability of the human brain): they are much more consistent in other species. Whereas access to more conventional 'gangs' are available to those with greater education, social resources, and cognitive abilities, it could be said that street gangs provide this need for those lacking such attributes, as well as opportunities for enrichment and living standards that are otherwise unattainable.[139] They are thus only one version of a widespread social feature of males, including but certainly not limited to humans, and part of their essential biological heritage. The traditions, structure, and culture of a given society will determine how this testosterone-driven behaviour asserts itself. This is an aspect not generally considered by those who study street gangs.

How far can we explain the apparently universal propensity of young men to form such groups? Before we discuss this further, we need to consider another associated feature of young men: their susceptibility to become fanatical. Loyalty to a group, tribe, gang or club is an essential element of membership and cohesion. This, in its most exaggerated form, manifests as excessive devotion or zeal to the group or its objectives, in some cases such that life itself, both of the person concerned and anyone he considers part of the 'outgroup', is willingly sacrificed. Most fanatics are young men.[140] Every despot, every dictator, knows this. The Japanese were able to recruit numerous kamikaze pilots in the Second World War: they were all young

men. There are relatively benign examples, such as football supporters, though violence between rival groups of these fanatics, a form of
war, is not unusual. Extreme adherence to political or religious causes
is a common basis for fanaticism. A charismatic (often older) cult
leader is often involved. Psychiatrists debate whether or not extreme
fanaticism, culminating in mass murder, qualifies as insanity: delusions are a symptom of mental illness (e.g. schizophrenia). But the
consensus is that, while murder may be the result of pathological
delusions, fanaticism itself (e.g. suicide bombing) is not. Social and
cultural milieus contribute: J. M. Post and colleagues write: 'Hope-
lessness, deprivation, envy, and humiliation make death, and paradise,
seem more appealing.'[141] Suicide bombers are typically young men
(late adolescents) who are uneducated, unmarried, and unemployed,
with low self-esteem but anxious for recognition and status by their
group (*amour propre*), though this is not always the case.[142] Adolescence is the time of life when parental bonds are loosened, and peer
groups become important. Recruitment into a group with focused
and powerful views (e.g. jihadists) is an empowering experience.[143]

It seems likely that the processes encouraging the formation of
gangs, and those that underlie fanaticism are closely related. The first
promotes the psychological need for the formation of groups with
common identities and bonds. The second cements those bonds by a
highly focused psychological state on the importance of membership
of the group and its objectives. We need to consider why it is that
these, in general, are a characteristic of young males.

Rupert Brooke, the poet, after enlisting, wrote to a friend:

> ...soldiering is the only life for me now. The training is a bloody bore.
> But on service one has a great feeling of fellowship, and a fine thrill, like
> nothing else in the world. And I'd not be able to exist for torment, if
> I weren't doing it. Not a bad place and time to die, Belgium in 1915? The
> world'll be tame enough after the war. For those who see it. Come and die.
> It'll be great fun.[144]

Which he did, though from a mosquito bite not a battle wound.

If this attitude is a consistent feature, then it is likely, in some way, to have advantages for survival, unattractive as that idea may seem. The young males of any group (and this applies to non-human primates as well as humans)[145] have two conflicting objectives. On the one hand, they must compete with each other for access to mates and all the other perquisites of social success. This, as we have seen, involves risk taking and potential danger, and is one feature of the behavioural actions of testosterone. But there is another requirement: the need for young males to bond together in such a way that concerted action is effective. Hunter-gatherers show this clearly: hunting a fleet and elusive prey is more easily done in cooperative groups. Cooperation is also essential for defence against other groups, or expansion of territory. High levels of risk-taking, in the interests of the group rather than the individual (resulting, in some cases, in fanaticism), are essential for effective cooperative action. Both individual and corporate risk-taking behaviour are ingredients of the way young males have to (and like to) behave. They do so because both types of behaviour have potent implications for survival, not only for them but also for their species or their nation. They have to be able to moderate each type of behaviour according to current social and physical demands.

As well as the imprint of biological inheritance, we see the tendrils of testosterone all over war, gangs, and fanaticism. Aggressive propensities, competitive traits, a tendency to form alliances with other group members but to be hostile towards strangers, territoriality: all are features of testosterone-driven behaviour, whether formed during exposure in early life or subsequently after puberty. The role of the brain in the impact that testosterone has on human life is considered more closely in another chapter (Chapter 10), but here we can consider those aspects relevant to war and the other manifestations described in this chapter. There are two possibilities to account for the roles of young men: either there is some second factor in addition to testosterone, particularly active in the young, that promotes inter-male bonding and all the other associated features of corporate aggression, or the state of the brain in young men enables testosterone to activate these behaviours.

Now is the moment to consider another hormone, which may influence the way testosterone functions (see also Chapter 4). If you watch someone milking a cow, you will see that he/she squeezes and stretches the cow's teats, and that this is rapidly followed by a squirt of milk. Human babies do the same to their mothers when they suckle: some women claim to be able to squirt milk across the room. This happens because stimulating the lactating nipple causes a pulse of the hormone oxytocin to be released from the pituitary gland. Breasts are sensitive to oxytocin, and some of their cells contract to express a shot of milk. Oxytocin is a small peptide—a chemical formed by a short string of amino acids joined together. Breasts and udders respond because their cells contain receptors to this peptide, which therefore acts as a signal for milk ejection. It is also important for the process of giving birth. For many years, this was thought to be what oxytocin did, so that its function in males—who also secrete oxytocin—remained mysterious. Then it was discovered that the brain contained oxytocin and its receptors, and that it could incite non-pregnant ewes, for example, to take an interest in lambs (which they otherwise ignore). So oxytocin had another, associated function: it was a powerful controller of maternal behaviour. It encouraged bonding between mother and infant. Oxytocin receptors in the brain increase during pregnancy, which was consistent with this role. Then it became clear that oxytocin was also concerned with other types of bonding.[146] Males of species of voles that formed relatively monogamous bonds with females had more oxytocin receptors than more promiscuous species.[147] Mating increases oxytocin, which in turn can stimulate sexual activity. Is oxytocin implicated in the tendency of males to bond and thus form gangs of various sorts?

There is a major problem about studying oxytocin in humans: small peptides such as oxytocin don't get into the brain when you inject them into the blood. One way around this is to spray them up the nose: some then enters the brain, though it's not a very consistent or dependable method. Males treated in this way show increased trust in others, become more willing to act together, and are more generous. In women at least, testosterone decreases trust. Men with low 2D:4D ratios (which may indicate higher exposure to prenatal testosterone: see Chapter 3) trust others less. Testosterone and oxytocin seem, to an extent, to be in opposition, which makes difficult any coherent idea of how they might combine to encourage group or gang formation. Oxytocin also improves conflict resolution, and increases preference for other members of the group (it may actually increase hostile reactions to outsiders). There is some evidence that it facilitates empathetic understanding of other peoples' feelings. You can see where this is leading. While we don't know why young men should be so vulnerable to bonding with each other in a common cause, it may be that young brains have more oxytocin, or receptors for it, than older ones. This leaves unanswered the question of how oxytocin might encourage bonding with other males in a social context, but with females in a sexual one. But we do know that the action of many neurochemicals on social behaviour depends on the context in which they are given.[148,149]

There are other features of the young male brain that might encourage uncritical bonding with others. As we discuss in more detail in Chapter 10 (and have noted in previous chapters), the frontal lobes of young men are not mature until their early 20s, and lag behind those of young women. The frontal lobes are concerned with many functions, including belief systems, social behaviour, and emotional responses. Is it too much to suggest that their immaturity may play a part in the willingness of young men to fight together, or to become members of a gang, regiment, football club and so on, and in some cases to be willing to sacrifice themselves? It could even be suggested that the delayed maturation of the frontal lobes of young men is an advantageous adaptation, in that it adds to their usefulness to their society by making them more liable to act riskily, aggressively, and in a war-like manner.

We can leave the philosophical, social, religious, political, and historical aspects of war to those qualified to discuss them. Testosterone does not cause war. War is the result of complex cognitive and emotional decisions made in specific cultural, political, and domestic circumstances (nearly always by males). But would war ever occur without men's testosterone? After all, wars are male events. Since if there were no testosterone then there would be no men (or women) this might seem an irrelevant question. Can we imagine a world in which testosterone retains its actions on sexuality, but not those perquisites of men that predispose to war? Also difficult to imagine is a world without competition, selection, and conflict over assets and mates. These are intrinsic to survival in the biological world, and hence these war-like actions are necessary for testosterone to be an effective mechanism for successful reproduction.

War can be biologically or socially advantageous for a victorious group or nation. For such a group to survive, its males have to be willing to mount a defence against attack by the males of another group. They also have to be prepared to try to increase their assets ('fitness') by capturing those of rival groups. Both attack and defence seem to be encouraged by behavioural tendencies clearly related to the actions of testosterone. So the occurrence of war seems to be an inevitable result of the powerful effects that testosterone has on male motivation, competitiveness, ambition, and risk-taking.

Wars have shaped our history, and continue to do so. What has altered in the modern world is not this basic propensity, but the technological and geopolitical contexts in which wars occur. This has meant that the evolution of the technology of war may have altered its biological usefulness. In those ancient times when war meant a raid on a neighbouring village by men armed with spears, the advantages gained—some food, a female or two, maybe the satisfaction of revenge—could be balanced against the possible loss or wounding of a few of the raiding party. The weapons and strategies generally did not lead to mass slaughter (though there were exceptions). Even if they did, this was on a local scale. As weapons and tactics developed, this changed. Armies became larger, wars became more protracted, killing became easier. Nations fought nations, coalitions of nations fought other coalitions. More sophisticated weapons (e.g. machine guns) caused massive casualties. Civilian casualties became greater (aerial bombing) with consequent ethical, emotional and economic consequences. Civilian casualties in the First World War accounted for about 10% of those killed; they were around 50% in the Second World War, only 20–25 years later (the UK figures are $c.0.2\%$ and $c.15\%$). Entire cities were devastated. The price to be paid became higher, the risk greater, the rewards less obvious. The objectives of conflict may be changing as well: as well as physical territory, control of information or culture may become more important. As wars became more complex, so too did their causes and consequences. The origins of the First World War, and whether it could have been avoided, are still hotly debated by historians. Wars on this scale had profound social, economic, and financial repercussions for the nation-states they involved in addition to the impact on the survival of individuals or their leaders. This has altered the balance between success and failure, and thus war as a strategy for biological and social rewards. Niall Ferguson writes:

> The victors of the First World War had paid a price far in excess of the value of all their gains; a price so high, indeed, that they would shortly find themselves quite unable to hold on to most of them…Quite apart from the killing, maiming and mourning, the war literally blew up the achievements of a century of economic advance…It was because of the war that [Hitler and Lenin] were able to rise to establish barbaric despotisms which perpetrated still more mass murder.

> Niall Ferguson (1998), *The Pity of War*. Basic Books, New York.

The title of Ferguson's book comes from a preface written by Wilfred Owen (himself killed in the First World War) and used on a commemorative tablet to war poets in Westminster Abbey: 'This book is not about heroes. English poetry is not yet fit to speak of them. / Nor is it about deeds, or lands, nor anything about glory, honour, might, majesty, dominion, or power, except War. / Above all I am not concerned with Poetry. My subject is War, and the pity of War. The Poetry is in the pity.' Both Owen and Brooke are among the names there.[150]

And so it was that, 25 years later, the Second World War broke out. The testosterone-driven inclination for war remains. This has had to be counteracted by cognitive political and moral restraints, themselves products of parts of the human brain that are not those directly sensitive to testosterone but regulate its actions (Chapter 10). Current events and recent history show that this control has been patchy, at best. We are still, it seems, trying to cope with the propensity for war in our world with brains evolved for a more simple and less technologically devastating one.

Testosterone is thus an essential contributor to the emergence of the war-like male responsible for the phenomenon of war.[151] There is little doubt that the effect it has on men's aggression, competitiveness, distrust of others, dominance-seeking, territoriality, and risk-appetite provides all the ingredients to make war-like behaviour highly probable; indeed, history suggests that testosterone-driven attributes in man have made war inevitable. If, that is, these tendencies are not moderated by other parts of the brain. But the ability of the social human brain to regulate the actions of testosterone and invent ways of gaining advantages and settling disputes by means other than collective conflict is very considerable, since it is these parts of the brain that are most highly developed in man.§ This implies that war is

§ The mass murder and chaos caused when the norms of a society break down is vividly illustrated by Ruth Scurr in her book *Fatal Purity: Robespierre and the French Revolution* (2007), Chatto and Windus.

not an inevitable feature of future history. How far war is contained depends, therefore, upon the analytical power of the recently (in archaeological terms) evolved areas of the human brain to make decisions about the costs and benefits of going to war, not of those more ancient ones sensitive to testosterone itself. These are discussed in more detail in Chapter 10. But the two work together: decisions about declaring war or taking part in it are made by these recently acquired (cortical) parts of the brain in leaders; similar areas may indicate restraint, or alternative strategies. The human brain has evolved methods of warfare that go beyond the use of weapons. For example, greater access to information may influence those called upon to fight, or support, such wars (e.g. encouraging group cohesion or demonizing an enemy: the 'narrative' of war[152]); but there are increasingly powerful methods of disseminating misinformation by those wishing to promote conflict. It is these newer areas of the brain that enable the extraordinary and unique technological evolution in humans. One consequence of this has been to alter the biological and social advantages of going to war. But deeper in the brain lie other, more ancient mechanisms, respondent to testosterone (and other hormones) that can bias such decisions one way or another. As in so many other contexts, at some point one has to consider the brain as a whole, rather than ascribing individual actions or outcomes to specific areas. Logic, memory, cognition, and emotion are intertwined in decisions to go to war. But testosterone lies at the core of this mix. We cannot always rely on ancient tendencies being moderated by more recently developed parts of the brain.[153]

At this point, let's recall the warning about thinking of testosterone as a metaphor for men. Sure, testosterone does all the male-typical things we have been discussing, and it's impossible to think of 'maleness' without testosterone. But testosterone has importance for women and so its significance is not limited to the consequences of what it does to men. Testosterone is not simply a 'male' hormone.

9

Testosterone in Women

How are we fallen! Fallen by mistaken rules,
And Education's more than Nature's fools;
Debarred from all improvements of the mind,
And to be dull, expected and designed;
And if someone would soar above the rest,
With warmer fancy, and ambition pressed,
So strong the opposing faction still appears,
The hopes to thrive can ne'er outweigh the fears.

Lady Winchelsea (c.1661), quoted by Virginia Woolf (1928),
A Room of One's Own. Penguin Books, London

I myself have never been able to find out precisely what feminism is. I only know that people call me a feminist when I express sentiments that differentiate me from a doormat.

Rebecca West quoted by Larry McMurtry,
New York Review of Books (2012)

Of the two sexes the woman is in the more powerful position. For the average woman is at the head of something with which she can do as she likes; the average man has to obey orders and do nothing else. He has to put one dull brick onto another dull brick.... The woman's world is a small one, perhaps, but she can alter it.... The average woman is a despot, the average man is a serf.

G. K. Chesterton (1958), 'Woman'. In: *Essays and Poems*,
W. Sheed (ed.). Penguin Books, Harmondsworth

So far, this book has been mostly about men. You might expect this in a book dedicated to testosterone. But here are a few surprises.

The first is that the blood of adult women before the menopause contains more testosterone than oestrogen (about five times). This, itself, does not necessarily mean that testosterone is the predominant hormone. As we saw in an earlier chapter (Chapter 2), the sensitivity of a hormone's receptor will determine how much hormone is needed for tissues to respond to it, and the oestrogen receptor is much more sensitive to oestrogen than the androgen receptor is to testosterone. So a simple direct comparison is not valid. Nevertheless, there is enough testosterone in a woman's blood (about one-tenth that of men) to activate her androgen receptors. Indeed, there are those who suggest that a woman's androgen receptors are more sensitive to testosterone than a man's. Testosterone levels in women tend to decline with age as in men, but, unlike men, individual differences in this decline are not associated with increasing ill-health.

The second surprise comes when we look a little more closely at the role a woman's hormones play in her sexuality. In female mammals such as rats or cats, oestrogens secreted from the ovaries have three functions. They enable fertility by triggering the female's pituitary gland to release another hormone (not a steroid, but a large peptide) that, in turn, triggers her ovary to release ripe eggs. Sometimes this occurs spontaneously, sometimes it is triggered by the act of mating (see Chapter 1). Oestrogen (sometimes together with progesterone) also makes her sexually active, so that she seeks out a male, or responds sexually to him (recall that females commonly select males: Chapter 6). Oestrogen has a third, equally important function: it makes the female periodically sexually attractive to males. The way this occurs varies between species. In some, oestrogen acts on the vagina to alter the scent it produces, and males find this attractive. A bitch 'on heat' thus attracts all the male dogs in the vicinity. In others, oestrogen may alter her appearance: in different species of monkey, females either grow rather huge swellings round their genitalia (some baboons), or the skin around their vagina changes colour (e.g. in rhesus monkeys, it turns bright red) around the middle of her cycle (when she is likely to be fertile); it also alters her smell. These signals are dependent on oestrogen. They are all signs to the male that the female is both fertile and sexually receptive; they also make her attractive. This coordinates sexual interaction, though, as previous chapters have related, much else needs to happen before a successful mating between two individuals can occur. Reproductive behaviour and fertility in these species are thus tightly controlled by hormones from the female's ovaries, mostly oestrogen with some help (in some species) from progesterone.

Things are different in women. Oestrogen plays a similar role in fertility to that in other mammals, by triggering the release of a monthly egg (ovulation; occasionally more than one egg) from her ovaries. It is also responsible for the considerable change in a woman's body at puberty, including the growth of her breasts. It acts on her vagina, making its lining thicker and able to secret moisture. But other functions are not the same as those in non-primate mammals. Reports of how a woman's sexuality changes during her menstrual cycle vary widely. Some report a marked peak in sexual interest near the middle of the cycle, when a woman is likely to be fertile; this is the pattern we might expect from what we know of other species. But others report differently: that they experience heightened sexual interest just before, or after, menstruation; or even that alterations in sexual desire have no clear relation to phases of their cycles. This has given rise to the idea that women are somehow 'emancipated' from their hormones, because of their large brains. But some women will tell you that they are only too aware of the phases of their menstrual cycle: mood and sexuality (they are obviously related) may show quite violent swings. It is more accurate to say that there is considerable individual variation in women's sensitivity and reactivity to the fluctuations of their ovarian hormones (oestrogen and progesterone) that occur during the menstrual cycle (Fig. 17).

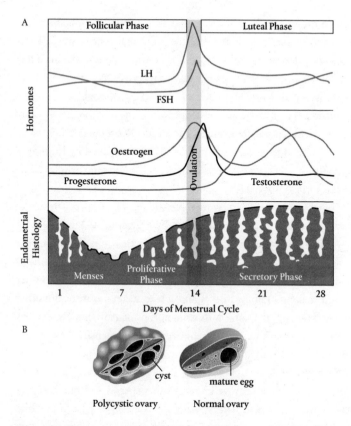

Fig. 17. (A) Variations in hormones during the menstrual cycle. Luteinizing hormone (LH) and Follicle-stimulating hormone (FSH) are the two gonadotrophins from the pituitary: they peak at mid-cycle, when ovulation occurs. LH is particularly important for ovulation. Oestrogen also peaks at mid-cycle, but progesterone is only secreted in large amounts during the second half of the cycle. Testosterone tends to peak at mid-cycle as well though some researchers have not confirmed this (there is probably considerable individual variation). The changes that occur in the lining of the uterus are shown at the bottom. (B) Normal and polycystic ovaries. The latter tend to secrete large amounts of testosterone.

There is another feature that distinguishes human females from those of other species. They don't advertise their fertility in the same way. Of course, wearing clothes might obstruct signals from the vagina or genital region, but there really aren't any to conceal. There has been much debate over the significance of the cryptic nature of the human female's reproductive strategy. Is this one way of ensuring that males stay around, since they will never know for sure when a female is fertile? Is this the reason behind the absence of marked oestrous periods in human females? There have been suggestions that persistent sexuality is the glue that binds couples together (Chapter 1), though they are currently decried as being too simple and lacking explanatory power for the complex structure of human societies. Bonding between human males and females, the argument goes, requires more than just sexual attraction or availability: necessary, perhaps, but not sufficient. We will have more to say about bonding, and the role of the brain, in Chapter 10.

The menopause is another human phenomenon. At around 50 years, a woman's ovaries run out of eggs, and also stop secreting both oestrogen and progesterone. If this were to happen in rats, cats, etc., all sexual activity would cease. This is not the case for women, though sexual activity may decline in some women (possibly the majority) after the menopause. There are several reasons for this: one is that the vagina is very sensitive to oestrogen, so it atrophies and becomes dry in its absence. This can result in sexual activity becoming painful, and hence aversive. Another is that ill-health becomes more common with ageing, and this will have an impact. An ageing or overfamiliar partner may be a factor. Side-effects of the menopause (e.g. hot flushes) may contribute. There may be alterations in mood though whether this is more likely after the menopause is disputed. At this point, we may be tempted to begin to discount any important role for hormones in the sexuality of human females. But there is a third surprise.

In the 1960s, reports appeared that removal of both the ovaries and adrenals of women (a rare but sometimes necessary event) resulted in marked decreases in their sexuality. The adrenals don't produce oestrogen, but they do secrete some testosterone. So do the ovaries. When these women were given small amounts of testosterone (about a fifth of that given to men), their sexuality improved. Experiments on female monkeys showed that testosterone acted on the brain (the hypothalamus) to stimulate sexual interest (sometimes termed 'libido' in humans).[154] There is now an established role for testosterone in the sexuality of women. So we can stop thinking about 'male' and 'female' hormones: testosterone is important for women and, as Chapter 2 describes, some of a man's testosterone is converted into oestrogen. Testosterone levels decline with age in some women, though the post-menopausal ovary may continue to secrete some testosterone. Treating those post-menopausal women who complain of reduced sexual interest with testosterone has consistently shown benefits, in some cases at least (about 50%).[155] The brains of women given testosterone show increased activity in response to the sight of male faces (Fig. 18). This makes it very strange that this is not a standard treatment for such complaints, and that a testosterone formulation for women is not available in most countries, including the UK. So women have to use one designed for men, though reducing the amount they take or apply to the skin.

Evidently, hormonal control of sexuality has not disappeared entirely in human females,[156] though there has been a curious shift in the way it occurs. What are we to make of this? There are two distinct sets of hormones in women: oestrogen, concerned with her sexual attractiveness (e.g. breast development), and testosterone, concerned with her sexual interest.* The two, as we have already seen,

---

* Some refer to this as sexual 'receptivity', a term really derived from studies on animals.

A

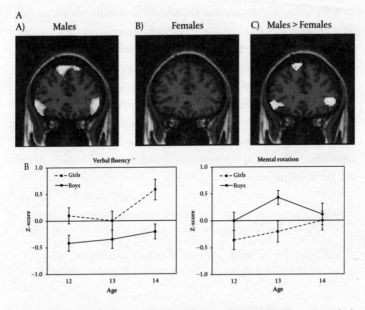

Fig. 18. (A) Sex differences in activation of the brain (fMRI) after subjects were asked to rate faces as socially approachable. Males showed greater activation than females in areas of the frontal cortex, though whether this indicates greater caution or increased distrust is not clear (see Chapter 10 for more discussion of this part of the brain). (B) Sex differences in verbal fluency (better in girls) and mental rotation (better in boys). These differences are present before puberty and do not change during it, suggesting that different hormone levels during adulthood are not implicated. The possible contribution of prenatal testosterone is discussed in the text.

are distinct aspects of sexual behaviour. Both are obviously essential for a satisfactory and functional sexual life. Although hormonal regulation of sexuality in women is still apparent, it has become divorced from fertility (which is heavily dependent on oestrogen). Since testosterone changes rather little during the menstrual cycle (but variably between women: in some there is a small mid-cycle peak), this may be one reason why marked alterations in sexuality are not consistent. We can only speculate on why this has happened: in non-primate species, a rather different arrangement serves them very well. But we have already remarked that reproductive strategies vary considerably between the females of even closely related species (Chapter 1). It does have implications for human sexuality: one is that the separation of sexual behaviour from fertility in females will reduce the likelihood compared to other species of a sexual act resulting in pregnancy. Does this imply that males need to maintain a long-term relation to ensure they have progeny; or that the fact that such relationships exist removes the need for a close link between sexuality and fertility? There are suggestions that such strategies are very different between the sexes:

> Whereas males fight for the right to fertilize as many females as possible, the situation for females is completely different. Whether she copulates with one or a hundred males, it will not alter the number of children she will give birth to. Jealousy among females is therefore less marked. Female competition occurs almost exclusively in the pair-bonded species, such as many birds and a few mammals. In those cases, females try to gain or defend a long-term tie with a male. Our own species is a good example: research by David Buss has demonstrated that whereas men get most upset at the thought of their wife or girlfriend having sex with another man, women dislike most the thought that their husband or boyfriend actually *loves* another woman, regardless of whether or not sex has occurred.   F. de Waal (1998), *Chimpanzee Politics*. Revised edition. Johns Hopkins University Press, Baltimore

But the conclusion must be that testosterone seems as important for sexual behaviour in human females as for males, which will surprise many.[†]

Does testosterone also have the pervasive effects on other aspects of women's behaviour and physiology that are so apparent in men? Without delving too deeply into the socio-political issues surrounding

---

[†] Post-menopausal hormone treatment is common. Tibolone, an artificial steroid that has some testosterone-like properties, is particularly effective in restoring or retaining post-menopausal women's sexual interest.

feminism, it should be noted that these questions have usually been asked in the context of masculinity. It is assumed that, if testosterone has such effects in women, they will be similar to those so well documented in men: heightened aggression, competitiveness, risk-taking, and so on. In other words, making them more 'male-like'. But let us now remember that testosterone is not an exclusively 'male' hormone, but has a real biological role in women. So this assumption may not be at all justified. There are some similarities. Female bodybuilders who take testosterone (or related steroids) show the same bizarre overdevelopment of their muscles as occurs in men. Women given testosterone show increased responses to social challenges, characteristic of men, but women with higher levels of testosterone actually show reduced aggressive responses to angry faces. Higher levels of testosterone predict less risk aversion in both men and women: it also predicts their choice of career—those with higher levels tended to choose risky jobs in finance.[157] The Iowa gambling task tests a propensity for risky economic decisions. Participants choose cards from four decks; some cards are rewarded, others penalized. The decks are stacked so that they offer different rates of success; some give smaller or larger gains over the long run; others losses. Subjects quickly realize which are the riskier packs (but may give bigger rewards). Higher levels of testosterone predict riskier choices in both sexes. Males usually show less empathy for others than females; testosterone reduced empathy in women. Neural responses to fear-inducing situations are reduced by testosterone in both sexes.[158]

There are some differences between men and women that can be counteracted, to a degree, by giving females testosterone: this is consistent with the idea that 'testosterone' equals 'male'. In other cases, this is not so. Testosterone seems to have similar actions in both sexes in some contexts; in others its effects on women are different from those in men.

There is a large literature on psychological differences between men and women.[159] These differences are mainly small, if they are reliable; they also overlap: there will be some men who are better than some women even in abilities that are generally (statistically) better in women. But most importantly, there is often little indication of whether these differences have real biological or social meaning; it is assumed, rather than demonstrated, that they do. Nevertheless, we do know that in many contexts, small differences in ability can have significant consequences on careers, social roles, and success or the unfolding of lifetime events. The Darwinian concept of evolution is based on the accumulative effect of small differences. So we should not discount the possibility that psychological sex differences demonstrated in the lab might have real life importance. Another question is whether these sex differences are dependent on testosterone. We cannot assume that sex differences in any psychological or behavioural ability, however robust, are necessarily the result of exposure (or lack of it) to testosterone either in early or adult life. There are many sex differences (e.g. in the presence of genes on the sex chromosomes) that may be independent of the actions of testosterone. So while we can catalogue these differences, their reliance or otherwise on testosterone has, in many cases, still to be established.

Some are the stuff of folklore and reliable material for stand-up comics. Women are notoriously poor at map-reading (we should point out here the considerable variability in both sexes: there are certainly women who are good, and men who are poor: this applies to all gender differences). But objective tests do confirm this: women, in general, perform less well than men on tasks that require the ability to navigate, or remember a route. Another similar task is called mental rotation: here the subject is shown a fairly complex shape, and then several views of the same shape from different angles. But some are actually not the same, and the subject has to rotate each image in his/her mind's eye to see if it matches the first one. It's a powerful test of visuo-spatial ability, and women (in general) are less good at it than men.[160] What does this mean, and why should it have occurred? One possible answer lies in the emergence of sex-differentiated roles early in evolution. Men hunt: they therefore need highly developed abilities in navigating through forest, etc., and remembering where prey, etc., are located. They also need to use their weapons accurately. They may have been selected for their visuo-spatial skills. Women are also less good at solving mathematical problems than men, though whether this represents intrinsic differences in ability, or the consequence of differences in education, is much debated.[161]

But women are better (in general) than men in their ability to recognize and remember faces, in correctly interpreting emotional expression on those faces, in empathy, and in several tests involving language (e.g. verbal fluency[162] (Fig. 18B)). In other words, women are better communicators and have greater empathy. The interpretation of this is obvious: the emerging role of females during evolution was in more domestic tasks: caring for the children, preparing (in some cases, gathering) the food, etc., requiring subtle and frequent communication with children and other women.[163] Higher skills in this area would be an advantage. Overall, there are no sex differences in cognitive ability between males and females, and such differences in particular abilities as occur seem to be present early in life,[164] suggesting they are determined not by adult levels of testosterone, but events occurring during development. This might include exposure (or lack of it) to testosterone in the womb, but this has still to be established. Let us, at this point, recall the important distinction between equality and similarity, already made in an earlier chapter.

Advances in technology have shown that some of the supposed differences in social roles between males and females are more to do with physical ability (i.e. strength) than anything else. During the Second World War, the development of machines enabled women to carry out tasks (e.g. making munitions) that were traditionally done by men. Power steering enables women to drive buses. Even more recently, the use of drones, for example, has led to women taking active roles in war that were unknown to previous generations.

Women athletes now compete with an enthusiasm comparable to men. Does their testosterone play a part? There is no doubt that taking testosterone or related compounds markedly improves their performance. Whether this is the result of increased muscular power, or heightened competitiveness ('the will to win') is not clear: it may be both. It seems that, in the case of female athletes, testosterone or its equivalents did make them more 'male-like' in the sense that the differences between their performance and that of men were reduced. But there is a limit to the information to be gained by taking excessive amounts of such steroids, which have (in both sexes) serious consequences for long-term mental and physical health. One result of this was that East German women athletes were later compensated for being duped into taking these drugs.

So could we abolish differences in gender by altering upbringing? Schemes exist to minimize gender-stereotypical play behaviour, for example in some Scandinavian nurseries. While this may have some impact, research on such children has nevertheless shown that little boys still prefer to play with trains, and little girls with dolls. Giving such toys to societies that have never seen them in real life has the same result. Furthermore, even little female monkeys prefer to investigate dolls, whereas males choose trucks.

We have already discussed how females exposed to abnormally high levels of testosterone during embryonic life may display more of the characteristics usually associated with males (Chapter 3). The 2D:4D ratio (Chapter 2) has been used as a proxy for early testosterone exposure (though recall the caveats about this measure). Women with low ratios (i.e. more resembling that of males) trust others less. Men trust others less than women (on the whole).[165] There seems no evidence that testosterone early in life in the female foetus plays any part in reducing characteristics we like to label as 'feminine', but it may bias the developing child towards a more masculine make-up.

One of the most dramatic examples of the way in which testoster-one-driven characteristics have been carried forward from primeval times to the present is in the world of finance (see Chapter 7 for more detailed discussion). Women play increasing, but still relatively minor, roles in finance. Studies on investment strategy by women show that they are less confident, more risk-averse, seek more advice, and have more modest views of their own competence than men.[166] There seems to be no firm evidence on whether the performance of female finance professionals is different from males. There is no evidence, either, on whether we can attribute differences in investment strategy to the (lack of) effects of either early (embryonic) or later exposure to testosterone, but the marked similarity between the features that characterize these gender differences and what we know of the effects of testosterone make such a speculation very appealing. We should recall here that the world of finance has been constructed by men, and is therefore likely to fit male attributes; female financiers are thus, for the present at least, working in a largely man-made environment.

What, then, can we say about the role of testosterone in women? On the one hand, there are instances in which testosterone exposure results in male-like characteristics. During embryonic life, it seems clear that testosterone has the role of encouraging male-type devel-opment, just as the absence of the ability to respond to testosterone (the androgen insensitivity syndrome) in a genetic male results in someone who is essentially female (Chapter 2). But once a female has been formed, the postnatal situation may be different. Although we should be very cautious about drawing parallels between rodent and human sexuality, evidence from the former shows that early exposure to testosterone not only 'masculinizes' the brain and other organs (e.g. the penis), it also increases the sensitivity of these organs to later testosterone secreted during adult life. If this is true for women, then testosterone in adulthood will act on parts of her body without the influence of previous exposure during embryonic life. This may have important results on the way testosterone functions in adult women. There are still circumstances in which testosterone administration (in excessive amounts) results in a masculine-type change: the example of female bodybuilders is one. But this might be behind the other instances in which testosterone has a purely female-type effect, such as its action on sexuality. Testosterone does not encourage adult females to behave sexually as males, or to find other females sexually attractive: rather, it has a physiological role in maintaining a component of female-type sexuality. This may apply to other aspects of female behaviour, though this is much less well-established. One example is that, contrary to its effects in men, it actually increased the level of 'fair' offers in the ultimatum game[167] (see Chapter 7 for an explanation of this game). The common idea that women who succeed in roles that are more often occupied by men must necessarily have more 'male' hormone (testosterone) is a fallacy, though a common one.[‡]

Polycystic ovarian syndrome[168] (PCOS) is a relatively common disorder in women. The ovaries produce excess amounts of testosterone (and other related steroids) (Fig. 17B). Why this happens is not known, though the syndrome seems to be strongly inherited, and so may be due to an (unknown) genetic variant. It is very often associated with insensitivity to insulin, which predisposes the woman to diabetes and obesity. The ovary may contain cysts (unruptured egg follicles) and also may not produce mature eggs, so affected women may not menstruate and can be infertile. One other effect of excess testosterone is that those affected can become very hirsute, to the extent that they may need to shave. Much of the research into this syndrome is focused on unravelling its cause, and improving treatment (which is quite

[‡] There is a well-known cartoon of Margaret Thatcher standing in a male urinal.

successful in many cases), but there are some studies on its other aspects. PCOS is associated with higher levels of anxiety and depression than in the general population. While this could be a direct effect of testosterone (though usually considered beneficial for depression) it is more likely to be secondary to concerns about personal appearance or infertility, or other aspects of the affected individual's quality of life. There are no reports of excessive sexual activity or desire, despite the established role of testosterone in human female sexual behaviour.[169] Neither does PCOS result in more muscular power; those with the syndrome don't look like weightlifters or body builders. In truth, PCOS has not told us very much about the role of testosterone in women, maybe because it is not a simple disorder, but complicated by the presence of abnormalities in insulin and related metabolic functions.

Sex differences in the brain have been much studied. There are several ways of doing this: looking at post-mortem brains, and staining them for the content of their neurochemicals or the size of various parts is one: another is to measure parts of the living brain using scanning or X-rays. Almost invariably, differences are found. The problem is to interpret them; that is, to map such differences in structure as are found on to what we know of the sex differences in cognitive or emotional function (which, as we need to keep remembering, may show significant differences, but with considerable overlap between the sexes). The most constant finding is that male brains are somewhat larger than females'. But since males are also larger, this isn't very helpful (elephants and whales have larger brains than humans). Furthermore, brain size changes with age, so it's important to be sure that comparisons take this into account.§ A recent detailed

§ Numerous attempts have been made to study the size of brains of so-called 'geniuses' in the belief that this might tell us why they have such gifts. All have proved worthless. A fascinating story of the study of Albert Einstein's brain (and the

analysis showed that some parts of the brain are larger in males, others in females.[170] In males, parts of the limbic system are commonly larger, which isn't obviously in line with the supposed greater empathetic abilities and emotionality of women (though we should also recall that measuring the volume or density of the brain has rather limited value). Areas larger in females include those concerned with language, which does make more sense.

An area of the hypothalamus, part of the brain well known to be concerned with sexuality (but also with eating, drinking, aggression and other means of survival: Chapter 10) is larger in males than females, and smaller in male-to-female transgender brains. What does this tell us? In fact, not very much, though it's intriguing. First, because the same finding has been reported in gay men compared to straight ones (so is it concerned with sexual preference?) and secondly, the size of a brain area tells us very little about what it does. Why should a larger INAH3 (the relevant area of the hypothalamus) alter any aspect of sexuality? The brain is not a muscle: what matters is the way it functions (that is, its connections and electrochemical activity). There have been reports of chemical differences in male and female brains, but no-one knows what these mean, or how (or whether) they relate to particular gender differences in brain function (and these are disputed). Scanning the brain has not helped much. There are different types of scan: the simplest tells us the size of different areas: as we've seen, not very useful. Alterations in the brain's blood flow in response to certain stimuli have been reported: but these often confirm what we already know—for example, that women and gay men respond differently to pictures of other men than do straight men. Connections between brain areas may differ: but we don't know how to interpret them in a very meaningful way. Areas larger in females include those

controversy this caused) is given by Michael Paterniti (2000) in his book: *Driving Mr Albert*. Abacus.

concerned with language, which does reflect ability. We need techniques that give us a more detailed and informative view of the brain's function. We also need to know whether testosterone plays an important role in sex differences in the developing or adult brain, and whether these are significant either for the normal function of the brain, or its susceptibility to disorder—which also shows marked sex differences: for example, depression is more common in women, but autism is more frequent in men.

So a difficult question is how behavioural differences between males and females are represented in the brain. Anatomical and neurochemical studies have revealed gender differences, but our current state of knowledge does not allow us to link these with behaviour. This is because the way the brain generates complex behaviour is not understood. I can't tell you exactly what happens in the brain when you feel hungry, or recognise a friend, or play a tennis stroke. Until we really understand such things, attempts to link gender differences in brain with those in behaviour remain obscure. We don't know what to look for. But it would be astounding if the different chromosomes (XX, XY) and hormones in males and females had no effect on the brain, and much scientific evidence shows that this does occur.

Recall, at this point, the evidence presented in Chapter 3 concerning the role of testosterone secreted during the first weeks of an embryo's life on its subsequent gender. Excessive testosterone, as in congenital adrenal hyperplasia (CAH) results in female babies whose genitals may look like those of males; conversely, if an XY individual (normally a male) has an androgen receptor that is non-functional (so the baby cannot respond to its own testosterone), then it grows up as a female, in most respects. So the absence of testosterone in a woman's life during her development in the womb is as important to her as its presence is to a male.

Because testosterone has been so associated with males, to the extent that it is seen as a metaphor for masculinity, its function in women has been considered uninteresting or even irrelevant. We know rather little about this hormone in women compared to the extensive literature on men. As this chapter shows, there are many questions still to be addressed, let alone answered. The only established role for testosterone in human females is in the context of sexuality. This should not be underestimated. But it is entirely possible that the reluctance of some, even those in the medical profession, to consider testosterone as contributing important qualities to a woman derives from the entrenched idea that testosterone belongs to males.

Why is there such difficulty in accepting that gender differences in the brain are one aspect of individuality? We readily agree that most men are stronger than most (but not all) women: there are good biological reasons for this. We also recognise that gender differences in behaviour or social roles are heavily influenced by current social rules and customs (just as a man's strength can be influenced by training etc.). So a particular woman's social role and her feeling of identity in that society will be influenced by her genes, her upbringing, the society in which she finds herself, and the opportunities it offers her. It's the same for men. Gender identity is how a person expresses themselves in that society. If a society represses expressions of sexuality or emphasizes gender differences in opportunities and roles, this will influence how women and men see themselves. The important point here is that gender identity is both 'biological' and 'social'. But none of these factors results in a simple binary division.

But differences in anything—brain, skin colour, gender—is not a justification for discrimination. To deny that gender differences in the brain exist (though they overlap) is to deny a basic aspect of sociobiology, a matter for celebration and wonder. To use these differences as a reason for subjugating women is indefensible. To repeat an earlier statement: there is an essential distinction to be made between similarity and equality. Denying the existence of gender differences in the brain confuses them, and the evidence against this view is overwhelming. The human brain has the cognitive ability to recognize inequality, decide that it may be unacceptable, and to formulate social and political mechanisms to reduce or eliminate it. But that doesn't imply the need to eliminate individuality, of which gender differences are a part.

Does the study of transgender people help us understand gender differences? Many of them report being convinced, from a very early age, that they were trapped in the 'wrong' body. So at an early stage in development, something has determined their gender identity, which is at variance with their body—in particular, their genitalia. Contemporary neuroscience is unable to give the answers we really need. We know a lot about the structure, arrangement, connections, and variations in individual nerve cells (neurons). We also know something about which part of the brain does what (see Chapter 10), though recent progress has been less. But the brain is not a pile of individual neurons. They form assemblies, and it's these assemblies that produce the functions we all know about: thoughts, emotions, perceptions, memories, self-esteem, and so on. Now here's the problem: we just don't know how neural assemblies do these things. There isn't even a generally accepted theory. Until we have one, we can't even speculate how the brain might encode gender identity or differences, along with all its other functions. So even if we find a plausible site in the brain that might be responsible for, say, gender identity, we wouldn't know how it did it. Therefore, we wouldn't know why it was that a person's idea of him/herself or their social role was at variance with their bodies. To describe something is not to explain it, though it may be the first step. We do know, from the recognition of the androgen insensitivity syndrome (see Chapter 3), that testosterone can play a major role in determining gender identity, one component of gender differences. But how does it do it? Those who take up vociferous positions on the social position of transgender people, or the best way to help them, should recognize the limits of their knowledge, and maybe adopt a more humble and tolerant point of view. There are many uncertainties, and plenty of topics rich for genuine disagreements.

We have only touched briefly so far on the role of testosterone in the organ encompassing our humanity and individuality. It is now time to consider what we know of the way testosterone acts on the brains of men (and women) to exert its remarkable and all-pervasive influence.

10

Testosterone and the Brain

The brain is waking and with it the mind is returning. It is as if the
Milky Way entered upon some cosmic dance. Swiftly the head-
mass becomes an enchanted loom where millions of flashing
shuttles weave a dissolving pattern, always a meaningful pattern
though never an abiding one; a shifting harmony of sub-patterns.

C. S. Sherrington (1940), *Man on His Nature*.
Penguin Books, London

We now use language, read books, watch TV, buy or grow most of
our food, occupy all continents and oceans, keep members of our
own and other species in cages, and are exterminating most other
animal and plant species, while the great apes still speechlessly
gather wild fruit in the jungle, occupy small ranges in the Old
World tropics, and threaten the existence of no other species.

Jared Diamond (1997), *Why Is Sex Fun?*
Weidenfeld & Nicolson, London.

No matter how sophisticated our science and technology, advanced our culture, or powerful our robotic auxiliaries, Homo
sapiens remains...a relatively unchanged biological species.
Therein lies our strength, and our weakness. It is the nature of all
biological species to multiply and expand heedlessly until the
environment bites back. The bite consists of feedback loops—
disease, famine, war and competition for scarce resources—
which intensify until pressure on the environment is eased. Add
to them the one feedback loop uniquely available to Homo sapiens
that can damp all the rest: conscious restraint.

E. O. Wilson (2002), *The Future of Life*.
Little, Brown, London.

It is his brain that makes a man, not his testosterone. No matter how large his muscles or how fertile his semen, if a male's behaviour isn't strongly adapted to reproductive success, then muscles and testes will go to waste.* We have seen that achieving this success requires a whole armoury of related behaviours and attributes, of which sexual motivation is an essential element, but only one of a necessary set, many of which depend on the brain. The brain is thus a major target of testosterone: how much do we know of what it does in the brain to produce the spectrum of masculine behaviour[171] that countless observations of both animals and humans have shown over many, many years?

We have also seen that testosterone is able to grow muscles and enlarge prostates and penises and so on because these tissues possess androgen receptors. The receptors both detect testosterone, and are the mechanism by which it regulates the tissues themselves. The brain is no different. Nerve cells (neurons) also make ('express') androgen receptors, which tells us that they use much the same mechanisms as other testosterone-sensitive tissues. But now for an important difference. Muscles and other testosterone-sensitive tissues are relatively simple, and most of their cells (but not necessarily all) possess androgen (testosterone) receptors. Many of these tissues respond to testosterone by increased growth of either the number of cells or their size. The adult brain has hardly any capacity for growing. Most adult neurons are incapable of replication (though there are exceptions in two or three areas of the brain). So how do brain cells react to testosterone?

As we saw in Chapter 2, once testosterone enters a nerve cell that has androgen receptors, it binds to these receptors and the whole complex then moves to attach itself to special regions of genes. These

---

* 'Natural selection doesn't give a fig for our happiness or sadness; brain mechanisms express these responses in whatever ways promote the long-term success of our genes.' R. N. Nesse (1999), 'The evolution of hope and despair'. *Social Research*, vol. 66, 429–69.

genes then are either activated or repressed. This will alter the proteins the nerve cell makes, and so change its function. There is a big problem: we have little idea of what many of these testosterone-sensitive genes do (and there are many) so we cannot, at the moment, explain how testosterone acts in the brain by its action on genes. The fact that the brain can convert testosterone to oestrogen, which acts on a different set of receptors, only complicates the picture. It can also convert testosterone to DHT (dihydrotestosterone), a metabolite that has even greater ability to bind to, and activate, the androgen receptor though this may be rather more important for tissues in the rest of the body than the brain. We know genes are important, if only because the effects of testosterone on sexual behaviour (see Chapter 4) are quite slow—this is what one would expect from altered gene activity. But there is another mechanism. Nerve cells, like all others, are contained in a bag-like membrane, which also covers the processes (dendrites) that project from every nerve cell. This membrane is covered in receptors, because most substances (steroids like testosterone are an exception) can't penetrate the cell membrane, and so have to act on the receptors in the membrane. These receptors play an essential role in the brain: nearly all transmitters (chemical links between nerve cells) relay information from one cell to another via receptors. That's how the brain works. As well as activating special receptors inside the cell (that is, inside the membrane), steroids such as testosterone may also act on a second set within the membrane itself. These are very different from those inside the cell. They act much faster, for example: whereas the classical steroid receptor may take hours to do anything, membrane receptors act in seconds or even milliseconds. This means that rapid changes in testosterone, such as those occurring after winning or defeat (Chapter 7) may also be able to alter brain activity or plasticity. So far, we have no clear evidence of the signi-ficance of such changes, though altered risk-taking under extreme conditions needing urgent decisions may be an example (Chapter 7).

The brain is, as we all know, highly complex, and has many functions. Is the whole brain sensitive to testosterone, or is the sensitivity restricted to certain specialized parts? If the latter were true, then it would give us one clue about how and where testosterone has its remarkable actions on behaviour. One way to approach the question is to map the distribution of androgen receptors in the brain. This can be done in various ways: one is to make an antibody to the receptor (which is a protein) and use this to locate the receptors. Since antibodies bind to their protein, if one marks the antibody in some way so as to make it visible, then one can see, under the microscope, where they are. Using this approach, it is immediately obvious that androgen receptors are present in certain areas of the brain, but completely absent in others. There are other ways of mapping androgen receptors in the brain, and they broadly give the same results. To appreciate what this map might mean, we need to digress for a moment, and discuss the way the brain is organized, and how this might relate to the behaviour we observe.

The idea that various parts of the brain have different functions has probably been around from ancient times, when early investigators first noticed (or recorded) that the brain wasn't a homogenous structure. But it was Thomas Willis, a seventeenth-century Oxford scientist, who first expressed the idea clearly, suggesting that different brain areas were responsible for distinct functions (e.g. 'thought' and 'action'), and thus heralding much of modern neuroscience. Though he made such fundamental contributions, Willis is known to most medical students today only because the arteries at the base of the brain are named after him ('circle of Willis'). But he did much more than that.[172] History can be cruel to a reputation.

In the last century, the American neuroscientist Paul MacLean, who was interested in comparative anatomy and the evolution of emotion, proposed a very simple system which he called the 'triune brain' (Fig. 19). This divided the brain into three parts: the most ancient was the 'reptilian complex', which was concerned with basic functions like

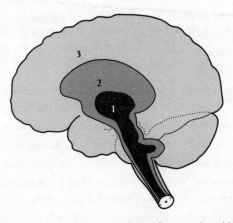

Fig. 19. Paul MacLean's 'triune' brain: (1) reptilian (reflex); (2) paleo (old)-mammalian (emotion); (3) neo (new)-mammalian (cognition).

breathing and movement; then came the 'limbic system', responsible for emotion, including experiencing pleasure and pain; finally the 'neocortex', which enables thought, reasoning and speech, and other 'higher' mental functions. Humans, he supposed, inherited all three parts—though the neocortex was markedly more complex than in other primates. These parts were sometimes in competition or even at odds with each other, an idea that has obvious resonance with notions that testosterone is part of an ancient hormonal mechanism regulating behaviour, but which has been carried forward as the human brain and behaviour evolved its complexity.[173]

> MacLean has shown that the R-complex [the subcortical forebrain systems including the limbic system] plays an important role in aggressive behavior, territoriality, ritual and the establishment of social hierarchies. Despite occasional welcome exceptions, this seems to me to characterize a great deal of modern human bureaucratic and political behavior. I do not mean that the neocortex is not functioning at all in American political convention or a meeting of the Supreme Soviet; after all, a great deal of the communication at such rituals is verbal and therefore neocortical. But

it is striking how much of our actual behavior—as distinguished from what we say and think about it—can be described in reptilian terms.

Carl Sagan (1977), *The Dragons of Eden*.
Hodder and Stoughton, London

MacLean's ideas have been criticized. Apart from being really too simple (it seems unlikely that one can classify the functions of the brain into only three categories), subsequent work has shown that the original boundaries proposed are less distinct than was once supposed; and, as we have seen, dividing 'emotion' from 'reason' is itself problematical when so much of what the brain does involves both. Nevertheless, let's focus on the limbic system.

Any animal including man has to survive if it is to reproduce. Survival entails matching the internal needs of the body (food, water, warmth, etc.) with external resources: that is, knowing what you need and where (and how) to find it. The world is a tough place: things that you need (food, water, etc.) are not always easy to come by, and others want what you may want. But some part of the brain has to tell you what is required for survival at any given moment. For example, when your blood sugar levels drop, signalling that you are running out of energy supply, the brain creates an experience we call 'hunger'. This has a double action: the brain makes the state itself unpleasant (being hungry distracts you from most other actions and you want to redress it), and it renders food highly attractive. You long to eat. This essential function involves several parts of the brain. If they don't work properly you starve, even in the midst of plenty. Similar mechanisms protect you against dehydration (thirst) or a drop in body temperature (seeking warmth), etc. Your brain makes you do things that are good for you by making you like them, and dislike the corresponding deficit state.

Not far behind your eyes, at the very base of the brain, there is a rather small area of the brain called the hypothalamus (Fig. 20A & B). It has already been mentioned in an earlier chapter. One of the functions of the hypothalamus is to monitor levels of essential signals in the

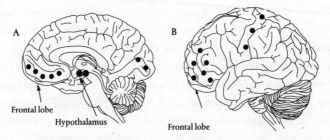

A B

Frontal lobe
Hypothalamus
Frontal lobe

Fig. 20. The parts of the human brain that contain androgen receptors (dark spots). (A) View of one half of the brain from the inside. In the cortex, receptors are concentrated in the lower part of the frontal lobe: the functions of this area are described in the text. There are a few at the back of the brain, a region that is concerned with vision. There are large concentrations in the hypothalamus. (B) The brain from the outside: receptors cluster in the frontal lobe, but there are a few in an area concerned with movement.

blood. For example, it measures blood sugar, but also your body temperature, as well as your water levels and blood pressure, and so on. You might compare it to the monitoring system in your house that measures temperature or the one in your car that tracks oil pressure. In order to maintain the delicate internal environment of the body within acceptable limits (a process termed 'homeostasis' by Walter Cannon[174]), your brain needs to know when any essential function is about to leave those limits. The hypothalamus constantly monitors your body state.

Plenty of evidence suggests that the hypothalamus responds after detecting a lack of food, water, etc., by making you want these things. It seems to do so by a system of chemical signals released by its nerve cells: most of these signals are peptides, which are essentially very small proteins. They act as a code. A protein is made up of a string of amino acids, and peptides can be just three amino acids long (though most are substantially longer than this). Damage to the hypothalamus disturbs the regulation of the body; exactly how depends on which part is involved. Damaging one part may make an animal enormously fat, because the detection system for matching food intake to fat stores is impaired. Nearby is an area that, if damaged, results in failure to respond to lack of food (anorexia). Other damage to nearby sites in the hypothalamus may alter water intake, or salt appetite (adequate amounts of salt are essential). The essential point is that 'wanting' something (food, water, etc.), otherwise called 'motivation', is set up by specific chemical signals in the hypothalamus which match the 'wanting' state with the detected deficit (sugar, fluid, and so on). But it doesn't stop there.

The hypothalamus, buried deep in the base of the brain, has no information about where to get what the body needs or how to go about getting it. Nor is it necessarily the part of the brain that makes being hungry or thirsty, etc., unpleasant and therefore a state you want to redress. The emotional response to a deficit state, which we label as 'unpleasant', or the 'pleasant' experience that results when that state is rectified (when a hungry man eats), seems to be the province of another part of the brain, not far from the hypothalamus. This part of the brain, the amygdala, lies deep within it on either side of the more central hypothalamus (Fig. 21). As expected, there are plentiful nervous connections between the two. Damaging the amygdala has very different results from damaging the hypothalamus. The amygdala seems to be concerned with the emotional reaction (liking or disliking) that is an essential part of the behavioural response to any deficit state. The amygdala has information not only from the hypothalamus (about the internal state of the body) but from the cerebral cortex (MacLean's third division: the neocortex), the great mantle that forms the wrinkled outside of the human brain, and which receives complex information about the environment via the senses and analyses that information, so you know where things are and what they are. For example, you recognize a tin labelled 'beans' as food; you know that a tap gives water and so on. Allowing external and internal information to match generates emotion, so a hungry man (low blood sugar) finds a plate of food attractive. A replete man does not—it may even seem rather off-putting. The amygdala does even more than this: it can react to external events (these are analysed by the cortex) that might be dangerous, generating (together with other parts of the brain) what we call 'fear'. Getting food, water, or a mate can be perilous. There are some scientists who think that the principal role of the amygdala is the induction of fear, but actually the amygdala has a wider function that includes fear or anxiety—both necessary for survival. It is also important for certain kinds of memory formation: particularly associations between some external signal and either reward or fear. This may involve altered connections, or gene expression. So Pavlov's dogs learned to associate the sound of a bell with food, and salivated in anticipation. School children hearing the dinner bell do likewise. A gun pointed towards you would make you fearful if you recognized it as a gun and knew what guns did.

You need more: a map of the world, so you can find things, and a memory of where they are and what they are (a tap gives water: where's the nearest tap?). The hippocampus lies just behind the amygdala, and is essential for forming memories, which includes the location of items that are important to you. It has been reported that London taxi drivers, who need to learn a multitude of routes, have particularly large hippocampi. They might have become taxi drivers because they had large hippocampi and hence were able to learn their way round London, or the training itself might have increased the size of their hippocampi. Intriguingly, the latter seems to be the case (an example of the plasticity of the brain). Furthermore, those that didn't show this effect were unable to pass the tough exam that London taxi drivers have to endure ('the knowledge').[175]

Part of the cerebral cortex, which lies near the midline of the brain, has a rather simpler structure than the rest: it's called the cingulate gyrus. It has close connections with the other parts of the limbic system and is called the limbic cortex. The hypothalamus, amygdala,

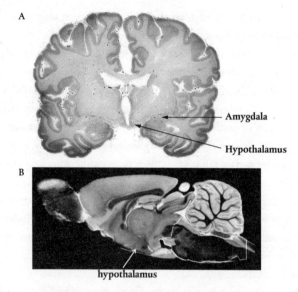

Fig. 21. (A) A section through the human brain: many receptors are present in the hypothalamus and amygdala, parts of the limbic system. (B) Longitudinal section through the rat's brain: the large protuberance at the front (left) is the olfactory bulb, emphasizing the importance of smell for rodents (in humans it's about the size of a small pea). The distribution of androgen receptors is very similar to that in the human brain.

hippocampus, and limbic cortex (and some other areas) together form the limbic system, which James Papez famously called the 'emotional brain' in 1937 (on very little evidence!)[176] (Fig. 21A & B). While there is no doubt that emotion is an important part of what this system does, it is not its only function. The whole concept of the limbic system has been questioned, but mainly on its definition as the 'emotional brain'. If one takes a more 'biological' view of what it does, recognizing that the limbic brain does more than 'emotion'—for example, regulating motivation and homeostasis—then these objections tend to disappear. However, we must always beware of drawing functional boundaries too distinctly round any part of the brain or any 'system';

they work together. The limbic system is more accurately called the 'survival' brain, because that reflects exactly what it does.[†] While one can find peptides all over the brain, they are particularly highly concentrated in the limbic system. This fact has suggested that, while the neurons of the cortex communicate with each other mainly by using rather simple chemicals (glutamate and ã-amino-butyric acid (GABA), which is a metabolite of glutamate), and rely on networks or assemblies of neurons for their complex functions, the limbic system relies more on the increased information encoded by its multiple peptides. They transmit a sort of neurochemical 'language'.

The reason that we have spent a little time on the limbic system and its function as an essential part of the brain's role in survival is that testosterone hijacks this system beautifully. It uses the limbic system to create emotional and motivational states that resemble, in many ways, those required for individual survival. But in this case the ultimate objective is not survival of the individual, but successful reproduction, and hence production of offspring and survival of the species. Testosterone—and reproduction in general—does require the individual to survive until he attains reproductive age and competence. It therefore contributes to that end (e.g. by making the individual competitive). But it is also concerned with survival of the species, in the sense that much of what testosterone does facilitates the ability of the individual male to pass on his genes to succeeding generations. So it needs to act on the brain in ways that ensure successful breeding, as well as aiding survival until breeding age. There are striking parallels between the way that hormones such as testosterone act on the brain, and those other chemical signals that initiate more immediate adaptations to want and need. Let's start with the hypothalamus.

[†] The 'survival brain' is described in more detail in my earlier book: J Herbert (2007), *The Minder Brain: How Your Brain Keeps You Alive, Protects You from Danger, and Ensures That You Reproduce*. World Scientific Press.

The hypothalamus (strictly speaking, the front part of it) is packed with androgen receptors. Testosterone clearly does something here. There are two traditional ways of finding out. The first is to destroy a tiny area of this part of the hypothalamus (a 'lesion'). This promptly stops a male rat (for example) from mating with a sexually attractive ('oestrous') female. Similar results are obtained with male guinea pigs, cats, even male monkeys. Giving any amount of testosterone doesn't restore mating after such a lesion. The male goes on looking uninterested. This doesn't prove that testosterone acts here, only that the area is essential for normal sexual behaviour. The second way is to castrate a male, allow his sexual activity to decline, and then see whether tiny implants of testosterone into the front of the hypothalamus—too small to act on other areas of the brain (one does control implants)—can restore the male's sexual interest. They do. But not always completely: because the amygdala has androgen receptors, so we may need implants here as well. In fact, there is a distributed network of androgen receptors, mainly (though not entirely) in the limbic system. The penis also needs testosterone, so giving testosterone only to the brain may result in sexual behaviour being incomplete, because the penis loses sensitivity or erectile ability. We do not have much direct evidence in humans, since we can't manipulate the human hypothalamus in this way (though rare lesions to the human hypothalamus have been observed to disrupt sexual behaviour), but we do know that the human hypothalamus has androgen receptors similar to those in monkeys and even rats. So testosterone acts on the limbic system in ways that are rather similar to those signals for food, water, salt, warmth, and so on; they make the individual 'want' sex (motivation) and 'like' females (emotional response): the amygdala is also packed with androgen receptors. The hippocampus has androgen receptors as well, though exactly what they do is still mysterious.

But there is still an unanswered question: what does testosterone do in the brain to cause a male to go trumpeting after females, despite possible threat to life or limb, in some species even ignoring food? Simply saying that it acts on receptors in the hypothalamus, amygdala, or wherever does not tell us much about *how* it acts. Somehow, testosterone makes the sight, smell, or sound of a fertile female attractive, and the act of mating rewarding. Neuroscientists have begun isolating a 'reward' system in the brain. This has focused attention on another region: the basal ganglia. These also lie deep in the brain, under the great cortical mantle. Traditionally, they have been associated with movement. Damage to the basal ganglia, or reduction in their content of dopamine (a chemical neurotransmitter), results in a variety of movement disorders, including Parkinson's disease. But movement is action, and action is caused by motivation. The general idea is that these ganglia (or part of them) enable a motivated action to take place by making it rewarding. They have plentiful connections with both the limbic system and region of the cortex (the frontal lobes) that regulate such behaviours. The basal ganglia have a rich supply of the transmitter dopamine, and reducing dopamine inhibits actions that would otherwise result in a reward of some kind; this includes sexual behaviour.

So far, so good. But there is a problem. Thirsty people seek water, not food, and vice versa. Rampant males may ignore both. Somewhere there must exist a way of making sure that what seems rewarding fits with the current state of the person (or animal); that is, matches the current biological or physiological need. The basal ganglia don't have many testosterone receptors, so it's unlikely that the decision to make sex rewarding lies here. But the reward system has to be biased according to circumstance. Testosterone points it towards sex. We don't know how this happens: it must occur in the limbic system (say, the amygdala, following information from the hypothalamus), but how this information is then transmitted to the reward system so that the things you want are the things you need, is still unclear. It may be that testosterone, contrary to what is usually believed, does not stimulate sexual motivation, but sets up a 'need' state (lack of sex), rather like lack of food or water induces corresponding need states. If so, the state itself would both activate and bias the reward system in the appropriate direction.

It has been repeatedly emphasized that, if testosterone is to enable successful reproduction, it has to do much more than activate sexuality. Competitiveness, aggression, risk-taking—these are just some of the qualities a successful male needs. And these are all attributes of the brain: how much do we know about how they come about?

Another way of investigating how the brain contributes to sex, aggression, or other testosterone-related behaviours is to see which parts are activated during these behaviours. Experimentally, there are several methods of doing this. One is to record the electrical activity of nerve cells (neurons). If this increases (or decreases) in particular parts of the brain, it is a clue. But very few neurons can be studied at any one time, and this limits the value of the technique. It is also quite difficult to interpret altered activity in terms of specific functions. And recording itself is difficult in freely moving animals who are actually behaving sexually (though it has been attempted). Another way is to look at genes. Neurons that are activated rapidly turn on particular genes (called 'immediate-early genes') that initiate alterations in their function. So by mapping the increased expression of these genes, one gets an idea of the pattern of neural activity in the brain. But this method also doesn't tell you what the neurons (or the genes) are doing, only that they are active. During sexual behaviour or exposure to the scent of a female, parts of the male's hypothalamus and amygdala increase the activation of their immediate early genes; a rather similar pattern emerges after an aggressive encounter. In the hypothalamus, there are both separate and overlapping neurons that respond to both, emphasizing the close relationship between sex and aggression. These results confirm the importance of the limbic system in these behaviours, but they do not tell us how these neurons generate sexual or aggressive behaviour. Measuring the amount of glucose they use is another way of assessing neural activity, with the same caveats. Looking at the pattern of genes in the brain that alter their expression (activity) after castration (and are also different between male and female brains) has revealed that there are a number (their functions vary widely).[177] This is fascinating, but efforts to pin a given gene or its protein product onto a given function of testosterone (e.g. aggression, sex, risk-taking, etc.) have not been very successful so far. So all these methods have been only partially useful in unravelling exactly what testosterone does in the brain, though each has contributed something. There is still a lot to be learned.

Even if we knew exactly what testosterone does to every neuron that contains an androgen receptor of some kind, would we know enough? While knowing more and more about how neurons work, and what affects their function, is hugely valuable, the brain is not simply a pile of neurons. They connect with each other, not randomly, but in certain patterns. These connections and patterns change under different circumstances. Neurons work as assemblies. We know that such assemblies can take on properties that cannot be predicted from the actions of individual neurons. It is these assemblies that construct desires and emotions, enable planning, and provide the knowledge of what to do in what circumstance. So if we are to really understand how testosterone has its manifold and pervasive actions on male behaviour, we need to understand how these assemblies actually work and what they represent. These are matters that reach beyond contemporary neuroscience, though everyone recognizes the objective. Bear this in mind as you read the rest of this chapter.

What about humans? None of the experimental methods can be used to study humans. In around 1965, Godfrey Hounsfield, a British electrical engineer, had an idea. He thought that he could reconstruct a three-dimensional model of an object (e.g. the brain) from multiple single X-rays ('slices') by computational methods. So CT (computed tomography) scanning was born; it continues to make huge contributions to clinical medicine, not only neurology.[178] It can show abnormalities in brain structure (e.g. the presence of a tumour, or dilated ventricles), but is less useful for studying the normal brain. There are other ways to image the living brain. Magnetic resonance imaging (MRI) uses radio waves rather than X-rays, and so is preferable for research (but is widely used clinically as well). MRI itself is rather like CT scanning, in that it shows the anatomy of the brain in three dimensions (both normal and abnormal). So one can measure the size of different areas of the brain. Functional MRI (fMRI) is different. Widely referred to as showing 'brain activity', in fact it depends on alterations in oxygenation of regions of the brain. Oxygenated blood enters each part of the brain through the local capillaries. Oxygen is extracted from the blood by the brain, and the deoxygenated blood passes into small veins, then larger ones, and so on, back to the heart. If a region of the brain is active, then—rather marvellously—the amount of blood going to it is increased (it needs more fuel). But this increase seems to be more than it needs, so some of the oxygenated blood spills over into the small veins. The brain blushes! This is picked up by the MRI (it's called the BOLD signal), and is what appears on the picture. So fMRI doesn't measure brain activity at all, but this overspill; from this one infers that the underlying brain is more active. The brain doesn't 'light up'; the blood vessels do. Although it is legitimate to infer than the underlying brain is more active, the exact relation between the BOLD signal and brain activity is still debated, as is the reason for the overflow of oxygenated blood. And the term 'brain activity' is not a very precise one. Claims that fMRI can 'see' your thoughts are simply untrue. There are other scanning methods, but fMRI has been the mainstay of research on the living human brain. Very useful, but—like all techniques—it has its limitations and should not be overinterpreted (but often is).

It is rather difficult to study human sexual behaviour in a scanner (though it has been attempted!). More usually, subjects are shown sexually explicit pictures. The pattern of fMRI response includes some limbic structures (e.g. hypothalamus, amygdala) and the limbic cortex, as one might expect (and hope), but other areas as well including those concerned with 'action' or 'reward' (the basal ganglia) and social control or decision-making (the frontal lobes). It is quite difficult to separate these effects from those really concerned with general arousal—which occurs after many stimuli, not just sexual ones—or the social environment. An interesting validation is to show either hetero- or homosexual men pictures of opposite or same sex individuals; their BOLD response is a highly reliable guide to their sexual orientation.[179] This method has been used as an attempt to diagnose other sexual preferences, including paedophilia. But there is no reliable information yet on whether testosterone alters the BOLD response in the brain to sexual stimuli, since studies on castrated men are rare, and those receiving 'anti-androgen' treatment are ill (e.g. have prostate cancer).

Brain damage in humans can result in increased aggression as well as uninhibited sexuality. And there have been plenty of studies on MRI scans in men or male adolescents with various forms of aggression, including impulsive (reactive) aggression and, in young people, conduct disorders and a psychological trait called 'callous-unemotionality', which refers to lack of sympathy or empathy for those suffering violence, and is often associated with high aggressive traits or conduct disorders. Interestingly, these studies point to three areas (there have been exceptions): the amygdala, the frontal cortex, and the front part of the cingulate cortex.[180] Let's consider each in turn.

The amygdala is the only one of the three to have plenty of andro-gen receptors. So if we are to look for an effect of testosterone, this is where we might start. Damage to the amygdala (which is rare) does alter aggressive behaviour,[‡] reactions to aggression (and other emotional events) are blunted, and both fear and aggression are reduced. During the last century, childhood aggressive or disruptive behaviour has been treated by making lesions in the amygdala[181] (this is no longer done). Adolescent and adult men with high levels of aggression may have smaller amygdalae than others. Those with genetic variations in serotonin genes that are associated with increased impulsivity and aggressive tendencies also show heightened BOLD responses in their amygdalae to aggressive stimuli. So far, so good. But where does testosterone fit in? We are looking not only for the activation of aggression in adults, but also for the formation of aggressive-type tendencies by testosterone early in life. Does testosterone in the amygdala result in men finding violence more attractive, or reacting violently to particular situations? Does testosterone mould the amygdala during embryonic life in this way? What we know is very limited and, to a degree, confusing. The volume of the amygdala is greater in boys than girls (allowing for brain size). Since boys are more aggressive than girls, this doesn't fit with the MRI finding of smaller amygdalae in aggressive boys. Higher testosterone levels in boys have also been associated with larger amygdalae: this agrees with the sex difference, but not the association of smaller amygdalae with aggressiveness. But CAH boys (who have been exposed to more testosterone during early life than normal, see Chapter 3) have smaller amygdalae and reduced memory for emotionally negative events.[182] This confused picture may tell us that simply measuring the volume of the

---

[‡] There is the rare Kluver–Bucy syndrome, first demonstrated in animals, that includes damage to the amygdala. A similar condition has occurred in humans. This is associated with indiscriminate sexual behaviour. Heinrich Kluver was a psychologist, Paul Bucy a neurosurgeon.

amygdala (or other parts of the brain) isn't always a very satisfactory method. We need to know how its structure, connections or chemical signals change. And we need to remember that the amygdala isn't only concerned with aggression, so correlating differences in its volume with particular patterns of behaviour is a problem.

A major theme of this book is that humans bring with them ancestral baggage in the form of basic brain mechanisms for sex, aggression, and other behaviours that represent their evolutionary history and have been essential for their success, but are now modulated by more recently developed parts of the brain[183] and the demands of a very different social and physical environment. It is these areas of the brain that are responsible for the complex and varied psychological, social, and cultural regulation of primeval human behaviour. The frontal lobes are an excellent example.

They lie, unsurprisingly, at the front of the brain and are particularly well developed in humans compared to other species. A gorilla brain looks very like a human one, yet the frontal lobes are obviously smaller (Chapter 1). Rats have hardly any. The term 'high-brow' refers to popular recognition that large frontal lobes and hence high foreheads signal high intelligence. They are responsible for such human-typical traits as planning, empathy, social awareness, decision-making, reward and emotional responses; all aspects of an individual's personality.

One of the brain's great mysteries is cognitive control. How does the brain produce behaviour that seems organized and willful?... cognitive control stems from patterns of activity in the prefrontal cortex that represent goals and the means to achieve them.... Virtually all complex behaviour involves constructing relationships between diverse, arbitrary pieces of information that have no intrinsic connection. Insight into the role of the prefrontal cortex in cognition can surely be gained from a better understanding of this process.

E. K. Miller, (2000), 'The prefrontal cortex and cognitive control'.
Nature Reviews Neuroscience 1, 59–65

There is some evidence that these functions can be assigned to different parts of the frontal lobes: in particular, the part lying just above the eyes (the 'orbito-frontal' lobe) seems important for emotion. No surprise, therefore, that it has major connections with the amygdala. The two work together.

Damage to the frontal lobes, which is not uncommon, causes major and often catastrophic changes in social functioning. The classic example, quoted everywhere, is Phineas Gage, a railway foreman of good and sober repute, who became degenerate, irresponsible, and socially and sexually inept after his frontal lobe had been damaged by a penetrating injury.[184] The frontal lobes are involved in much more than sexual and aggressive behaviour; they moderate practically all emotional and cognitive actions. For example, damage to the orbito-frontal lobes disturbs not only emotional responses, it also impairs financial decision-making (it enhances inappropriate risk-taking: Chapter 7), another example of the interplay between emotion and 'rational' thinking. Relatives often say that frontal lobe damage has resulted in a person different from the one they knew. More pertinently, is testosterone involved in any of these functions of the frontal lobes? There are few androgen receptors in this part of the brain.

Adolescence is a time of enormous emotional and behavioural turbulence. Not only is it characterized by greatly increased sexual activity, it is also marked—particularly in males—by heightened aggressiveness, deviant behaviour, risk-taking, and criminality. Recall that most violent crimes are committed by young males (Chapter 5). The frontal lobes are the last part of the brain to mature. They do this by altering both the number of cells ('grey matter') and fibres ('white matter') of which they are made. Two interesting facts: this process continues until late adolescence or even early adulthood (the rest of the brain has finished maturing); and it is slower in males than females. So adolescent males have an adult limbic and hormonal system, but immature frontal lobes. Can we ascribe the risk-taking, lack of social control, emotional instability of the typical adolescent male to this mismatch? It seems plausible (this is also considered in previous chapters). Whether the sex difference in frontal lobe development is the result of current testosterone, or their exposure to it in early life (or even something else) has not been clarified so far, but it does seem likely. Criminally violent behaviour has been correlated with underdeveloped or damaged frontal lobes in some cases (any such association is probabilistic rather than determinant).[185] So this area of the brain is involved in activities we associate with adolescence. We have to wonder why it is that delayed frontal lobe maturation has been apparently preserved in the process of natural selection. It might be that the trade-off between these seemingly anti-social behaviours and the costs or benefits they bring to society are actually positive on the whole. As we have seen, the risky and competitive behaviour of young males is an essential part of their biological fitness, unpleasant as it may seem to the rest of us from time to time.

The frontal part of the cingulate cortex is, traditionally, part of the 'limbic' cortex: so-called because its structure is somewhat simpler than other areas of the neocortex. Rather ignored for years, it has become a recent focus of research. It lies just behind the frontal cortex and was part of the original 'emotional brain' circuit described by James Papez. It has connections with both the hippocampus and the frontal cortex. It seems to be involved in both emotion and cognition, maybe as one link between the two. This might be a crucial function for the actions of testosterone. Both sexual and aggressive stimuli can activate it, though also many other areas of the cortex. Most evidence about the role of the cingulate cortex has come from MRI studies, with their attendant limitations. Much of the focus on it has concerned its role in depression and/or anxiety, but it does seem to play a part in empathetic responses to distress in others. All these conditions will have powerful secondary actions on sexuality and all its component behaviours.

The realization that individual differences in aggression, and hence in antisocial or even criminal behaviour, may be related to corresponding differences in the brain has led to the emergence of a new discipline: neurocriminology.[186] This attempts to attribute the risk of a particular individual (man) to behave criminally to predictive features in either his brain or his genetic makeup (the two are related). Important legal and ethical questions are raised: if one does identify such a genetic predisposition or brain state, then is the person responsible for his criminal behaviour? Should certain individuals be screened for such risk factors? If they are found, then what should be done? Are attempts to rectify or counteract such risk factors justified, and in what circumstances? Individual differences in testosterone itself have not been reliably identified as a risk factor for criminality (but, as we have seen, the majority of violent crimes are committed by young men), though other physiological features (such as a low heart rate, or low cortisol levels) have. Genetic predispositions have been said to account for about half the risk, though there are strong caveats to this statement. No single gene has been identified as carrying more than a very small increased risk (see the discussion of XYY men: Chapter 5); there is often confusion between categories of crime (e.g. white collar crime such as fraud and violent crimes: the two are very different); and genetic make-up may influence the quality of upbringing or other early adversities that might actually be responsible for later criminal tendencies. Variations in the androgen receptor have still to be studied. Brain imaging has shown some differences in the brain between those convicted of crime and controls, but one has to bear in mind that, if people behave differently, then this will certainly be reflected in particular patterns of brain activity. For example, reduced functioning (fMRI) of the frontal lobes may represent the reason, at a neural level, why some men are more likely to commit crimes than others, but it may also reflect that fact. However, it seems true that damage to the frontal lobes predicts increased likelihood of criminal behaviour—

particularly damage to the orbitofrontal cortex, known to be concerned with emotional responses and decision-making (see earlier in the chapter). This suggests that it is the control of testosterone-driven behaviour, rather than the effect of testosterone itself, which might be responsible (see Chapter 6 for additional discussion of this topic). But although scanning may show us that there are differences in the brain of those who do, or do not, commit crime, it isn't clear what these differences mean in terms of altered function or why they should predict criminality.

The distinguishing feature of the human brain is the great development of the cerebral cortex, the wrinkled structure that forms so much of the two cerebral hemispheres, and which gives the human brain its characteristic appearance. The wrinkling is the result of this great development: to retain a head that is a reasonable size, the cerebral cortex (actually a large thick sheet of neural tissue) is crammed into the skull, much as one crams one's handkerchief into a pocket. The fact that the wrinkles are more or less the same (though only approximately) between individuals is simply because the 'cramming'—that is, the growth of the brain within the skull—occurs in much the same way in all of us.

Nervous systems that are hard-wired are lightweight, energy-efficient, and fine for organisms that cope with stereotyped environments on a limited budget. Fancier brains, thanks to their plasticity, are capable not just of stereotyped anticipation, but also of adjusting to trends ... For truly high-powered control, what you want is an anticipation machine that will adjust itself in major ways in a few milliseconds, and for that you need a virtuoso future-producer, a system that can think ahead, avoid ruts in its own activity, solve problems in advance of encountering them, and recognize entirely novel harbingers of good and ill. For all our foolishness, we human beings are vastly better equipped for that task than any other self-controllers, and it is our enormous brains that make this possible.   Daniel C. Dennett (1991), *Consciousness Explained.*
                                            Little, Brown and Co. New York

What we know about the way that testosterone acts on the brain reinforces the central premise of this book. The sites of androgen receptors in the brain, which are relatively constant across all species of male mammals so far studied, including humans, points clearly to a common neural mechanism by which testosterone activates sexual behaviour and the other primeval behaviours associated with successful reproduction. The hypothalamus, amygdala and other parts of the limbic system are constructions of the human brain as they are of others. But there is a great mass of human brain, not necessarily responsive to testosterone, but evolved for wider purposes, that regulates, steers, and moderates sexual behaviour as well as the expression of other, associated, behaviours. We only need to acknowledge the powerful control that society exerts over sexuality in so many ways through religion, customs, social hierarchy, and the rest. We do not need in this book to discuss them in detail (there are plenty of books that do), simply to point to their existence and the fact that they vary so widely across cultures and across time to be sure that it is not the limbic brain, which remains so similar in many species, but the new (neo) cortex, so prominently flowering in recent evolutionary history in primates and most of all in man, that is responsible. This is the part of the brain that allows such flexibility in the detail of control of the actions of testosterone, always so powerful in execution. Other species have controls, as we have seen, but none so varied and various as ours. Testosterone has little say in this, but everything it does is subservient to the control exerted by the more recently evolved parts of the human brain. Yet, without the basic mechanism of testosterone, without its primal activation of sexuality and all the associated behaviours and bodily functions, no amount of higher cortical activity would enable successful reproduction. Just as we would struggle to draw clear boundaries around the different areas of the brain, so their functions are intertwined.

Deconstructing the brain is essential if we want to understand how different parts of it contribute to its function. But the brain, in the end, works as whole. It is an indisputable fact that within each of our skulls there lies around 1.5 kg of an extraordinary tissue, an organ responsible for our history, our present, and the future of our descendants. The brain is what makes a human being. This is not an original, or even a new idea: it was evident to a great mind 2,500 years ago:

> Men ought to know that from the brain and from the brain only arise our pleasures, joys, laughter and jests as well as our sorrows, pains, griefs and tears ... It is the same thing which makes us mad or delirious, inspires us with dread and fear, whether by night or by day, brings sleeplessness, inopportune mistakes, aimless anxieties, absent mindedness and acts that are contrary to habit ....
>
> Hippocrates, *c.*400 BC, quoted by E. S. Valenstein (1973), *Brain Control*. John Wiley, Chichester

Thoughts, dreams, memories, ambitions, love, our knowledge of the world gained by individual experience, the transmission of more knowledge through language and records made by those in other places or from other times, what we see, feel, do, and think: all this we do with our brain and more: including consciousness, that quality known to us all, but still retaining much of its mystery despite centuries of debate by philosophers, and more recent attention from neuroscientists: the crowning gift that this most complex, most wonderful, organ bestows on us. So when it malfunctions, we are devastated. I would not wish to overstate the case: our history is not all testosterone-dependent. But the influence of testosterone, ancient but fundamental for our survival, permeates much of what we do. Even though human life is so varied and our technological world is so complex, testosterone goes on playing its ancient role. But in modern man it acts by channelling and controlling our complex brains during the course of a huge variety of behaviours. It pits individuals against each other in competition in every aspect of life; it has driven wars; it may lie behind much creative endeavour—and of course, without it, there would be no human beings. A human being, by which I mean a human brain, is not made by testosterone; it is made by human genes but shaped by human experience.

> Humans and chimps share 98.4 percent of their DNA, slightly more than either does with the gorilla. The orang-utan is less closely related and the New World monkeys even less so. Any idea that humans are on a lofty genetic pinnacle is simply wrong. A taxonomist from Mars armed with a DNA hybridisation machine would classify humans, gorillas and chimpanzees as members of the same closely-related biological family.... Our brains and our behaviour are what separate humans from any other animal. They probably involve a few genes whose importance is lost in a measure of genetic difference.... a whole set of intellectual and cultural attributes ... appear once a crucial level of intelligence has been reached and which are not coded for by genes at all.
>
> Steve Jones (1993), *The Language of the Genes*. HarperCollins, London

But to understand our humanity, which is to understand our brains, we need to take a cool look at testosterone.

Written by a Victorian genius, in comfortable, ornate but modest prose, rich with learning and generous references to friends and colleagues, the *Origin of Species* is the most influential book on science ever published. Charles Darwin, though he knew well the function of the testes, had no knowledge of testosterone. But as you read these lines, consider how far the remarkable role of testosterone, celebrated in this book, might have played in what Darwin wrote so long ago:

> When we no longer look at an organic being as a savage looks at a ship ... when we regard every production of nature as one which has had a history; when we contemplate every complex structure and instinct as the summing up of many contrivances, each useful to the possessor, nearly in the same way as when we look at any great mechanical invention as the summing up of the labour, the experience, the reason, and

even the blunders of numerous workmen; when we thus view each organic being, how far more interesting...will the study of natural history become!   Charles Darwin (1859), *The Origin of Species*

Testosterone really does shape a man's history, for without it he could never become a man. Then it shapes what he looks like, how he behaves, whether he has children, and much of the sort of life he leads. But generalities disguise individuality. Variations, both genetic and circumstantial, in the amount of testosterone or when it is secreted, and how each male responds to his own testosterone, determine many of the differences that distinguish one man from another (Fig. 22). Testosterone guides the way that a man interacts with others, and so impacts on his social history. Within that society there will be leaders and followers, successes and failures: the story of testosterone tells us that it makes powerful contributions to such outcomes. Competitiveness has other functions, including driving ambition and inventiveness and thus innovation and technical advance. But a man's history is not only his own, but that of his ancestors; and they, too,

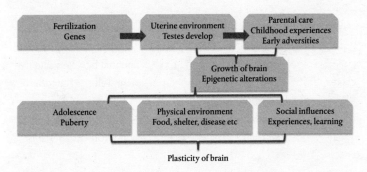

Fig. 22. There are manifold influences on the brain during life, from the earliest moment of fertilization, during embryonic life, childhood, and the slings and arrows of adulthood. Testosterone is only one of them. Occasionally, this process results in a genius. What distinguishes geniuses from the rest of us is not their testosterone but their brains. But contemporary neuroscience is not at a stage that allows it to tell us what the essential difference might be.

were influenced by testosterone in similar ways. Just as it is difficult to predict the behaviour of an assembly of neurons from that of individual cells, so a society takes on features of its own, determined, in some complex way, by the behaviour and propensities of individuals. Since history is social as well as personal, the roles of competition, and even war, in the unfolding of the history of a town, a district, a nation along time, with all its triumphs and disasters, is influenced by this simple but wondrous chemical. And the human brain has had to devise multiple ways of regulating, channelling, and optimizing the powerful effects of testosterone on male behaviour through laws, religion, and customs. So testosterone lies at the heart of much of our history. Its effects are not confined to men: the lives of women, because of their interaction with men, are shaped by it, and by their own testosterone. But at the end of any discussion of the impact of testosterone on the history of mankind in all its wide-reaching and powerful complexity, a simple fact remains: without testosterone there would be no humans to have a history.

NOTES

1. Some Eskimos apparently build a nightly igloo, as they have done for centuries: but the tools they now use to do so have, presumably, improved.

2. It has to be said that Charles Darwin, in *The Descent of Man, and Selection in Relation to Sex* stressed similarities, rather than the differences, between human and animal brains and behaviour. But he was making the case for a common origin and the process of evolution.

3. Technologies may take on properties other than their original one: for example, the form of buildings may be expressions of art or sacred beliefs as well as utility; this is another property of the human brain. See J. Rykwert (1963), *The Idea of a Town: the Anthropology of Urban Form in Rome, Italy and the Ancient World*.

4. For example, there are considerable, but unexplained, cross-species variations in the way the placenta is formed. See Y. W. Loke (2013), *Life's Vital Link: The Astonishing Role of the Placenta*. Oxford University Press.

5. 'The development of a larger neocortex has enabled motivated behavior to occur such that maternal affiliation may take place without pregnancy and parturition. This unique development in human evolution has matched parturient females with non-parturient females in sustaining the behavioral potential for infant caregiving. Decreasing the dependence of maternal behavior on endocrine determinants has been an evolutionary necessity in order for infant care to extend beyond the influence of pregnancy or suckling hormones. In most mammals, maternal care ceases when suckling terminates.' E. B. Keverne (2001), 'Genomic imprinting, maternal care, and brain evolution'. *Hormones and Behavior*, vol. 40, pp. 146–55.

6. Vividly displayed in the film *Ratatouille* (2007, Pixar Animation).

7. A comprehensive account of the variety of human sexual behaviour and how it has changed in different cultures and periods is given by P. B. Gray and J. R. Garcia (2013) in their book *Evolution and Human Sexual Behavior* (Harvard University Press). Their preface also has a useful bibliography of books on human and primate sexuality.

8. Charles Darwin (1872), *The Origin of Species* (sixth edition), R. E. Leakey (ed.). Hill and Wang, New York.

9. Derived from Greek 'gone' (gonads: testis or ovary) and 'trophe' (nourishes). The placenta also produces a similar hormone.

10. They are called Leydig cells. There is a special gene (called DMRT1) that makes these cells what they are.

11. So the Second World War ditty sung about Hitler by British soldiers was not actually the insult that was intended.

12. An analogy: a radio station transmits a message, but only those with a radio tuned to the right frequency will be able to detect (hear) it.

13. C. Lovatt Evans (1964), *Reminiscences of Bayliss and Starling*. Lecture delivered before the Physiological Society, 1963.

14. W. Meuser and E. Nieshlag (1977), 'Sex hormones and depth of voice in the male'. *Deutsche Medizin Wochenshrift*, vol. 102, pp. 261–4.

15. Melissa Hines (2011), *Brain Gender*. Oxford University Press.

16. D. G. Zuloaga, D. A. Puts, C. L. Jordan and S. M. Breedlove (2008), 'The role of androgen receptors in the masculinization of brain and behavior: what we've learned from the testicular feminization mutation'. *Hormones and Behavior*, vol. 53, pp. 613–26.

17. There are those who dispute this, and report that the male testes continue to secrete testosterone throughout embryonic life. See P. Sarkar et al. (2007), 'Amniotic fluid testosterone: relationship with cortisol and gestational age'. *Clinical Endocrinology*, vol. 67, pp. 743–7.

18. For example, John Colapinto (2000), *As Nature Made Him: The Boy Who Was Raised As A Girl*. HarperCollins, New York.

19. S. J. Bradley et al. (1998), 'Experiment of nurture: ablatio penis at 2 months, and a psychosexual follow-up in young adulthood'. *Pediatrics*, vol. 102, pp. e9.

20. W. G. Reiner (2001), 'Gender identity and sex-of-rearing in children with disorders of sexual differentiation'. *Journal of Pediatric Endocrinology and Metabolism*, vol. 189, pp. 549–53.

21. Melissa Hines (2011), *Brain Gender*. Oxford University Press.

22. More than a thousand different mutations have been described in the androgen receptor. Most are very rare. B. Gottleib et al. (1012), 'The androgen receptor gene mutations database: 2012 update'. *Human Mutation*, vol. 33, pp. 887–94.

23. H. J. Kang et al. (2013), 'The effect of 5á-reductase-2 deficiency on human fertility'. *Fertility and Sterility*, vol. 101, pp. 310–16.

24. Jeremy Bentham (1785), *Offences Against One's Self: Paederasty*.

25. 'Several experimental and clinical findings suggest that androgen deficiency in genetic males...when occurring during a critical period of brain differentiation can predispose to...homosexuality...if these findings could be confirmed...it might become possible in the future...to correct abnormal sex hormone levels...in order to prevent...homosexuality.' G. Dorner (1983), *Archives of Sexual Behavior*, vol. 12, p. 1.

26. '...a substantial minority, perhaps a third, (of male-to-female transgenders) are not sexually attracted to male partners. A much smaller number of female-to-male transgenders are sexually attracted to male partners.' R. Green (2009), *New Oxford Textbook of Psychiatry*, M. Gelder et al. (ed.). Oxford, p. 844.

27. R. Green (1987), *The 'Sissy Boy Syndrome' and the Development of Homosexuality*. Yale University Press.

28. Serotonin may play a role in determining sexual preference. S. Zhang et al. (2013), 'Serotonin signaling in the brain of adult female mice is required for sexual preference'. *Proceedings of the National Academy of Sciences USA*, vol. 110, pp. 9968–73.

29. P. L. Hurd, K. L. Vaillancourt, and N. L. Dinsdale (2011), *Behavioral Genetics*, vol. 41, pp. 543–56.

30. 'The many reports so far make it clear that there is some relationship between digit ratios and sexually differentiated human behavior. Those relationships can be most parsimoniously explained by a masculinizing effect of prenatal androgens on both digits and behavior.' S. M. Breedlove (2010), *Endocrinology*, vol. 151, pp. 4116–22.

31. M.-C. Lai, M. V. Lombardo, and S. Baron-Cohen (2013), 'Autism'. *Lancet*, vol. 383, pp. 896–91.

32. R. Knickmeyer et al. (2006), 'Androgens and autistic traits: A study of individuals with congenital adrenal hyperplasia'. *Hormones and Behavior*, vol. 50, pp. 148–53.

33. Testosterone levels are reported to be higher in the mothers of autistic children than controls, suggesting another possible mechanism. X.-J. Xu et al. (2013), 'Mothers of autistic children: lower plasma levels of oxytocin and arg-vasopressin and a higher level of testosterone', *PLoS. One*, vol. 8, e74849.

34. Kazio Ishiguro, in his novel *The Remains of the Day*, illustrates this well. The protagonist is a butler whose ostensibly simple life is shaped by his employer's disastrous involvement in events preceding the Second World War. Only at the end of the book does he realize what has been happening to him: 'Lord Darlington wasn't a bad man. He wasn't a bad man at all. And at

least he had the privilege of being able to say at the end of his life that he made his own mistakes. His lordship was a courageous man. He chose a certain path in life, it proved to be a misguided one, but there, he chose it, he can say that at least. As for myself, I cannot even claim that. You see I *trusted*. I trusted his lordship's wisdom. All those years I served him. I trusted I was doing something worthwhile. I can't even say I made my own mistakes. Really—one has to ask oneself—what dignity is there in that?' K. Ishiguro (1989), *The Remains of the Day*. Faber and Faber, London.

35. This only applies, of course, to species living some way from the equator, where seasonal changes are prominent.

36. 'Attitude surveys in a wide variety of human societies around the world have shown that men tend to be more interested than women in sexual variety, including casual sex and brief relationships. That attitude is readily understandable because it tends to maximize transmission of the genes of a man but not of a woman.' Jared Diamond (1997), *Why Is Sex Fun?* Weidenfeld & Nicolson, London.

37. 'We believe that a deep comprehensive understanding of human sexual behavior develops most surely from a correct view of our species' place in the animal kingdom. Acquiring this view requires comparative study. In addition, there is substantial intrinsic interest in the sexual behavior of nonhuman species and its determinants. . . . Sexual processes are virtually ubiquitous in the animal kingdom. Sexual and evolutionary processes are so closely linked as to be almost inseparable. To understand evolution one must understand sex, and vice versa.' G. Bermant and J. M. Davidson (1974), *Biological Bases of Sexual Behavior*. Harper and Row, New York.

38. J. Bancroft (1989), *Human Sexuality and its Problems*. Churchill Livingstone, Edinburgh.

39. M. G. Forest et al. (1974), 'Hypophyso-gonadal function in humans during the first year of life. 1. Evidence for testicular activity in early infancy'. *Journal of Clinical Investigation*, vol. 53, pp. 819–28.

40. A. F. Dixson (1993), 'Effects of testosterone propionate upon the sexual and aggressive behavior of adult male marmosets (Callithrix jacchus) castrated as neonates'. *Hormones and Behavior*, vol. 27, pp. 216–30.

41. G. M. Alexander (2014), 'Postnatal testosterone concentrations and male social development'. *Frontiers in Endocrinology*, vol. 5, pp. 1–6.

42. J. Bancroft (2005), 'The endocrinology of sexual arousal'. *Journal of Endocrinology*, vol. 186, pp. 411–27.

43. Mating between pairs that have more traits in common than would be expected by chance.

44. Or man, in homosexual relationships.

45. 'Still, who could say what men ever were looking for? They looked for what they found; they knew what pleased them only when they saw it. No theory was valid in such matters, and nothing was more unaccountable or more natural than anything else.' Henry James (1881), *The Portrait of a Lady*.

46. For example, the work of Helen Fisher and colleagues: see 'Romantic love: a mammalian brain system for mate choice'. *Philosophical Transactions of the Royal Society* (2006), series B, vol. 361, pp. 2173–86.

47. 'Something in the way she moves / Attracts me like no other lover / Something in the way she woos me / I don't want to leave her now / You know I believe her now.' The Beatles: 'Something'.

48. 'Falling in love has often been regarded as the supreme adventure, the supreme romantic accident. In so much as there is in it something outside ourselves, something of a sort of merry fatalism, this is very true. Love does take us and transfigure and torture us. It does break our hearts with an unbearable beauty, like the unbearable beauty of music. But in so far as we have certainly something to do with the matter; in so far as we are in some sense prepared to fall in love and in some sense to jump into it; in so far as we do to some extent choose and to some extent even judge—in all this falling in love is not truly romantic, it is not truly adventurous at all. . . . The supreme adventure is not falling in love. The supreme adventure is being born.' G. K. Chesterton (1958), 'On certain modern writers and the institution of the family'. In: *Essays and Poems*, W. Sheed (ed.). Penguin Books, Harmondsworth.

49. Prescriptions for testosterone in ageing males have increased by 90% over the last 10 years or so. E. H. Gan et al. (2013), 'A UK epidemic of testosterone prescribing'. *Clinical Endocrinology*, vol. 79, pp. 564–70.

50. 'Generally, the most vigorous males, those which are best fitted for their places in nature, will leave most progeny. But in many cases victory depends not so much on general vigour as on having special weapons confined to the male sex. A hornless stag or spurless cock would have a poor chance of leaving numerous offspring. Sexual selection, by always allowing the victor to breed, might surely give indomitable courage, length to the spur, and strength to the wing to strike with the spurred leg, in nearly the same manner as does the brutal cockfighter by the careful selection of his best cocks.' Charles Darwin (1872), *The Origin of Species* (sixth edition), edited by R. E. Leakey. Hill and Wang, New York.

51. 'Of course Freud's immediate reception in America was not auspicious. A few professional alienists (an early term for a psychiatrist) understood his importance, but to most of the public he appeared as some kind of German sexologist, an exponent of free love who used big words to talk about dirty

things. At least a decade would pass before Freud would have his revenge and see his ideas begin to destroy sex in America forever.' E. L. Doctorow (1976), *Ragtime*. Pan Books, London.

52. C. D. Knott, T. M. Emery, R. M. Stumpf, and M. H. McIntyre (2010), 'Female reproductive strategies in orangutans, evidence for female choice and counterstrategies to infanticide in a species with frequent sexual coercion'. *Proceedings of the Royal Society B*, vol. 277, pp. 105–13.

53. G. A. Lincoln, F. Guinness, and R. V. Short (1972), 'The way in which testosterone controls the social and sexual behavior of the red deer stag (cervus elaphus)'. *Hormones and Behavior*, vol. 3, pp. 375–96.

54. S. Jones (2002), *Y: The Descent of Man*. Little Brown, London.

55. M. Potts and T. Hayden (2008), *Sex and War: How Biology Explains Warfare and Terrorism and Offers a Path to a Safer World*. BenBella Book, Dallas, TX.

56. Thomas Hobbes (1651), *Leviathan*.

57. Konrad Lorenz (1966), *On Aggression*. Methuen, London.

58. For example, D. S. Lehrman (1953), 'A critique of Konrad Lorenz's theory of instinctive behavior'. *Quarterly Review of Biology*, vol. 28, pp. 337–63, and the collection of essays in A. Montague (ed.) (1973), *Man and Aggression*. Oxford University Press.

59. Anthony Clare (2000), *Of Men: Masculinity in Crisis*. Chatto and Windus, London.

60. For a detailed discussion of the problem of defining aggression in humans, see M. Martinez and C. Blasco-Ros (2005), 'Typology of human aggression and its biological control'. In: Molecular mechanisms influencing aggressive behaviour. *Novartis Foundation Symposium*, vol. 268, pp. 201–15, Wiley, Chichester.

61. C. Levi-Strauss (1962), *The Savage Mind*. Weidenfeld & Nicolson, London.

62. A. F. Dixson (1998), *Primate Sexuality*. Oxford University Press.

63. M. Hines (2011), 'Gender development and the human brain'. *Annual Review of Neuroscience*, vol. 34, pp. 69–88.

64. W. I. Wong et al. (2013), 'Are there parental socialization effects on the sex-typical behavior of individual with congenital adrenal hyperplasia?' *Archives of Sexual Behavior*, vol. 42, pp. 381–91.

65. G. A. Mathews et al. (2009), 'Personality and congenital adrenal hyperplasia: possible effects of prenatal androgen exposure'. *Hormones and Behavior*, vol. 55, pp. 285–91.

66. Crime Survey of England and Wales 2011/12.

67. J. M. Carre et al. (2010), 'Motivational and situational factors and the relationship between testosterone dynamics and human aggression during competition'. *Biological Psychology*, vol. 84, pp. 346–53.

68. J. Klinesmith et al. (2006), 'Guns, testosterone, and aggression: an experimental test of a meditational hypothesis'. *Psychological Science*, vol. 17, pp. 568–71.

69. J. M. Carre et al. (2011), 'The social neuroendocrinology of human aggression'. *Psychoneuroendocrinology*, vol. 36, pp. 935–44.

70. J. C. Dreher et al. (2016), 'Testosterone causes both prosocial and anti-social status-enhancing behaviours in human males'. *Proceedings of the National Academy, USA* vol. 113, pp. 11633–11638.

71. R. Lovell-Badge (2005), 'Aggressive behaviour: contributions from genes on the Y chromosome'. *Novartis Foundation Symposium*, vol. 268, pp. 20–33. Wiley, Chichester.

72. A. Caspi et al. (2002), 'Role of genotype in the cycle of violence in maltreated children'. *Science*, vol. 297, pp. 851–4.

73. R. R. Thompson et al. (2006), 'Sex-specific influences of vasopressin on human social communication'. *Proceedings of the National Academy of Sciences*, vol. 103, pp. 7889–94.

74. P. H. Mehta and R. A. Josephs (2010), 'Testosterone and cortisol jointly regulate dominance: evidence for a dual-hormone hypothesis'. *Hormones and Behavior*, vol. 58, pp. 898–906.

75. T. R. Gurr (1981), 'Historical trends in violent crime: a critical review of the evidence'. *Crime and Justice*, vol. 3, pp. 295–353.

76. UN Office on Drugs and Crime.

77. Surveys consistently show that more men than women would prefer (and have had) multiple sexual partners. Cross-cultural comparisons also show that while polygamy is (and has been) not uncommon, polyandry is rare.

78. Thomas Hobbes (1651), *Leviathan*.

79. F. de Waal (1998), *Chimpanzee Politics. Power and Sex among Apes*. Revised edition. Johns Hopkins University Press, Baltimore.

80. If the recessive gene is advantageous, then inbreeding has beneficial results, as in thoroughbreds.

81. For example, consanguineous unions and the burden of disability: a population-based study in communities of Northeastern Brazil. M. Weller et al. (2012), *American Journal of Human Biology*, vol. 34, p. 835. They found a rate of about 3% of disabled infants for unrelated unions, but around 10% for related ones.

82. D. Lieberman et al. (2007), 'The architecture of human kin detection'. *Nature*, vol. 445, pp. 727–31.

83. Peter Brown, in his review of the book *From Shame to Sin: the Christian Transformation of Sexual Morality in Late Antiquity* by Kyle Harper (*New York Review of Books*, vol. 50, pp. 48–52) discusses the prevalence of presumably

enforced sex by slavery in Roman times, and the revolution brought by the emergence of Christian-based morality. Sex with slaves was also part of other cultures, including the United States during the eighteenth and nineteenth centuries. This illustrates both circumstances under which the normal control of sex by females could be overturned, and also the powerful influence of social mores in human beings (and thus the human brain) on sexual behaviour in general.

84. 'What do you want me to do? He said. / I want you to be considerate of a young girl's reputation. / I never meant not to be. / She smiled. I believe you, she said. But you must understand. This is another country. Here a woman's reputation is all she has. / Yes mam. / There is no forgiveness, you see. / Mam? / There is no forgiveness. For women. A man may lose his honor and regain it again. But a woman cannot. She cannot.' C. McCarthy (1993), *All the Pretty Horses*. Vintage International.

85. A recent comprehensive account is given by J. Bourke (2007), *Rape: A History from 1860 to the Present*. Virago, London.

86. I am not including what has been termed 'sexual coercion', and which includes some of the ways used by males to limit the sexual activities of others, or restrict those of females, described in the first part of this chapter. See M. N. Muller and R. W. Wrangham (eds) (2009), *Sexual Coercion in Primates and Humans*. Harvard University Press, Cambridge, MA.

87. M. P. McCabe and M. Wauchope (2005), 'Behavioral characteristics of men accused of rape: evidence for different types of rapists'. *Archives of Sexual Behavior*, vol. 34, pp. 241–53. The authors were able to divide men convicted of rape into four personality types: anger rapists, power rapists, exploitative rapists and sadistic rapists.

88. There have been disparate findings: Giotakos et al. report that testosterone levels were higher in rapists than a control group. Sex hormones and biogenic amine turnover of sex offenders in relation to their temperament and character dimensions. *Psychiatry Research*, vol. 127, pp. 185–93, but this is denied by Aromaki et al. (2000), 'Testosterone, sexuality and anti-social personality in rapists and child molesters: a pilot study'. *Psychiatry Research*, vol. 110, pp. 239–47.

89. S. Rajender et al. (2008), 'Reduced CAG repeats length in androgen receptor gene is associated with violent criminal behavior'. *International Journal of Legal Medicine*, vol. 122, pp. 367–72.

90. Crime survey for England and Wales, UK Government 2012.

91. UN Office on Drugs and Crime.

92. National Violence Against Women Survey, USA.

93. R. Jewkes et al. (2013), 'Prevalence of and factors associated with non-partner rape perpetration: findings from the UN multi-country cross-sectional study on men and violence in Asia and the Pacific'. *Lancet Global Health*, vol. 1, pp. e208–18.

94. In Papua New Guinea, it is sometime acceptable for a man to rape the wife of a less dominant one who cannot defend her. The rapist may then give a present of a pig as reparation. P. Matthiessen (1962), *Under the Mountain Wall*. Viking Press, New York.

95. A. Beevor (2002), *Berlin. The Downfall 1945*. Viking Books, London.

96. The Rape of Nanking in 1937 is another infamous example.

97. I. Buruma (2013), *Year Zero: a History of 1945*. The Penguin Press.

98. M. J. Escasa et al. (2011), 'Salivary testosterone levels in men at a U.S sex club'. *Archives of Sexual Behavior*, vol. 40, pp. 921–6.

99. But attempts have been made to represent life as a game. See, for example, John Conway's 'Game of Life'.

100. J. M. Carre and S. K. Punam (2010), 'Watching a previous victory produces an increase in testosterone in elite hockey players'. *Psychoneuroendocrinology*, vol. 35, pp. 475–9.

101. P. A. Brennan et al. (2011), 'Serum testosterone levels in surgeons during major head and neck surgery: a suppositional study'. *British Journal of Oral and Maxillofacial Surgery*, vol. 49, pp. 190–3.

102. S. J. Stanton et al. (2009), 'Dominance, politics, and physiology: voters' testosterone changes on the night of the 2008 United States presidential election'. *PLoS ONE*, vol. 4, e7543.

103. C. L. Apicella et al. (2014), 'Salivary testosterone change following monetary wins and losses predicts future financial risk-taking'. *Psychoneuroendocrinology*, vol. 39, pp. 58–64.

104. R. M. Rose et al. (1969), 'Androgen response to stress. II. Excretion of testosterone, epitestosterone, androsterone and etiocholanolone during basic combat training and under threat of attack'. *Psychosomatic Medicine*, vol. 31, pp. 418–36.

105. A. Alvergne et al. (2009), 'Variation in testosterone levels and male reproductive effort: insight from a polygynous human population'. *Hormones and Behavior*, vol. 56, pp. 491–7.

106. M. Hasegawa et al. (2008), 'Changes in salivary physiological stress markers associated with winning and losing'. *Biomedical Research*, vol. 29, pp. 43–6.

107. An entertaining and informative account of economic theory, financial decision-making and the activities of traders is given in M. Fenton-O'Creevy et al. (2005), *Traders. Risks, Decisions, and Management in Financial Markets*. Oxford University Press.

108. There is evidence, for example, that genes associated with serotonin can influence risk appetite and financial choices. C. M. Kuhnen et al. (2013), 'Serotoninergic genotypes, neuroticism, and financial choices'. *PLoS ONE*, vol. 8, e54632.

109. P. Xu et al. (2013), 'Neural basis of emotional decision making in trait anxiety'. *Journal of Neuroscience*, online 20 November 2013.

110. R. J. Dolan (2007), 'The human amygdala and orbital frontal cortex in behavioural regulation'. *Philosophical Transactions of the Royal Society*, vol. 362, pp. 787–99; E. Fehr and C. F. Camerer (2007), 'Social neuroeconomics: the neural circuitry of social preferences'. *Trends in Cognitive Sciences*, vol. 11, pp. 419–27.

111. For example, P. Slovic et al. (2004), 'Risk as analysis and risk as feelings: some thoughts about affect, reason, risk and rationality'. *Risk Analysis*, vol. 24, pp. 311–32; D. Kahneman (2011), *Thinking, Fast and Slow*. Macmillan, New York. The American psychologist Robert Zajonc was an early and influential exponent of the role of emotion in decision-making. See R. B. Zajonc (1980), 'Feeling and thinking: preferences need no inferences'. *American Psychologist*, vol. 35, pp. 151–75.

112. M. Baddeley (2010), 'Herding, social influence and economic decision-making: socio-psychological and neuroscientific analyses'. *Philosophical Transactions of the Royal Society*, vol. 365, pp. 281–90.

113. A comprehensive and highly readable account of economic theory is given by H.-J. Chang (2014), *Economics: The Users Guide*. Pelican Books.

114. C. F. Camerer (2008), 'Neuroeconomics: opening the gray box'. *Neuron*, vol. 60, pp. 416–19.

115. J. M. Coates and J. Herbert (2008), 'Endogenous steroids and financial risk taking on a London trading floor'. *Proceedings of the National Academy of Sciences, USA*, vol. 105, pp. 6167–72.

116. E. Garbarino et al. (2011), 'Digit ratios (2D:4D) as predictors of risky decision making for both sexes'. *Journal of Risk and Uncertainty*, vol. 42, pp. 1–26.

117. P. Branas-Garza and A. Rusticini (2011), 'Organizing effects of testosterone and economic behavior: not just risk taking'. *PLoS ONE*, vol. 6, e29842.

118. Recently, we have found that giving either testosterone or cortisol increases risk-taking in a game that models a trading floor. C. Cueva, R. E. Roberts et al. (2015) Cortisol and testosterone increase financial risk taking and may destabilise markets. Scientific Reports vol. 5, pp. 11–206 DOI: 10.1038/strep11206.

119. P. W. Zak et al. (2009), 'Testosterone administration decreases generosity in the ultimatum game'. *PLoS. ONE*, vol. 4, e8330.

120. N. D. Wright et al. (2012), 'Testosterone disrupts human collaboration by increasing egocentric choices'. *Proceedings of the Royal Society*, vol. 279, pp. 2275–80.

121. For example, W. Eckhardt (1992), *Civilizations, Empires and Wars: A Quantitative History of War*. MacFarland; M. Potts and T. Hayden (2008), *Sex and War: How Biology Explains Warfare and Terrorism and Offers a Path to a Safer World*. BenBella Books, Texas; K. L. Vaux (1992), *Ethics and the Gulf War: Religion, Rhetoric, and Righteousness*. Westview Press, Boulder; K. N. Waltz (2001), *Man, the State and War. A Theoretical Analysis*. Columbia University Press; N. Ferguson (1998), *The Pity of War*. Basic Books. See also endnote 35.

122. There are exceptions. Patricia Churchland, for example, has written books that attempt to relate philosophy with neuroscience, such as *Neurophilosophy: Towards a Unified Science of the Mind-Brain* (1986), 'MIT Press, and *Braintrust: What Neuroscience Tells Us About Morality* (2011), 'Princeton University Press.

123. C. von Clausewitz (1832–34), *On War*, originally *Vom Kriege*, 3 vols. Berlin.

124. A vivid account of the raids made by chimpanzees on others is given by R. Wrangham and D. Peterson (1996), *Demonic Males: Apes and the Origins of Human Violence*. Houghton Mifflin.

125. V. C. Wynne-Edwards (1986), *Evolution Through Group Selection*. Oxford: Blackwell Scientific.

126. For example, E. F. M. Durbin and J. Bowlby (1938), 'Personal aggressiveness and war'. Reprinted in *War. Studies from Psychology, Sociology, Anthropology*, L. Bramson and G. W. Goethals (eds) (1964), Basic Books, New York.

127. R. W. Wrangham and L. Glowacki (2012), 'Intergroup aggression in chimpanzees and war in nomadic hunter-gatherers'. *Human Nature*, vol. 32, pp. 5–29.

128. E. C. Tolman (1942), *Drives towards War*. Appleton-Century-Crofts, New York.

129. C. Hammer (2012), 'Why do soldiers fight?' *Historically Speaking*, vol. 13, pp. 10–12.

130. Boudica, queen of the Iceni, is an obvious exception. She seems to have been extraordinarily cruel.

131. 'Everywhere, all in our generation, and Pao and I with them, caught in the whirlpool of war. Some plunging into the current, joyous, eager for the testing, shouting of flag and country and cause, dying for symbols. Others, as young but unillusioned, seeing clearly the wasted sacrifice of life and the greater death in the hardening of spirits to hatred, lying and killing. We are all swirled into the current, rushed into experiences of terror and exaltation. There is no true adventure save within oneself. No experience has significance until one has received it and made it a part of oneself in thought.' Han Suyin (1994), *Destination Chungking*. Penguin Books, London.

132. N. Chagnon (1968), *Yanomamo: The Fierce People*. Holt Reinhart, Winston NY; but see L. E. Sponsel (1998), 'Yanomamo: an area of conflict and aggression in the Amazon'. *Aggressive Behavior*, vol. 24, pp. 97–122.

133. Dress is used widely in many societies, both advanced and less developed, to indicate status or rank.

134. 'There is no love-broker in the world can more prevail in man's commendation with women than report of valour.' William Shakespeare, *Twelfth Night*, act 2 scene 3.

135. R. W. Wrangham and M. L. Wilson (2004), 'Collective violence: comparisons between youths and chimpanzees'. *Annals of NY. Academy of Sciences*, vol. 1036, pp. 233–56.

136. Jared Diamond's book *Guns, Germs and Steel: The Fates of Human Societies* (1997, Norton, NY) is a clear explanation of the role of technology (but also many other factors) in social and political domination.

137. M. W. Klein (1995), *The American Street Gang*. Oxford University Press, New York.

138. The Cambridge Apostles (named after the original number) is another example. An undergraduate club, it was secretive, exclusive and aspirants underwent an elaborate vetting process. Status within the club was based not on fighting prowess, but on ability in academic debate (a form of competition). The Apostles became notorious when several of their members were revealed later (1950–60s) to be Soviet KGB spies. Similar rituals are seen in American university fraternities, and in the process of gaining 'manhood' in aboriginal tribes.

139. It is significant that entry to the clubs or fraternities formed by even educated young men commonly involve ritual humiliation, thus emphasizing the dominant status of the existing members. D. C. Brotherton (2015) *Youth street gangs. A critical appraisal*. Routledge, London and New York; J. A. Densley (2012) 'The organisation of London's street gangs'. *Global Crime*, vol. 13 pp. 42–64.

140. But there are exceptions. For example, the German Bader–Meinhoff gang included several core women members.

141. J. M. Post et al. (2009), 'The psychology of suicide terrorism'. *Psychiatry*, vol. 72, pp. 13–31.

142. R. Sela-Shayovitz (2007), 'Suicide bombers in Israel: their motivation, characteristics, and prior activity in terrorist organizations'. *International Journal of Conflict and Violence*, vol. 1, pp. 160–8. There were also female suicide bombers in the Tamil Tigers of Sri Lanka (who seem to have invented this strategy).

143. Sigmund Freud (1922, *Beyond the Pleasure Principle*) postulated a 'death instinct' (or 'death drive') which, he thought, was an 'urge to restore an

earlier state of things'; however, he seems to have had his own doubts about this idea.

144. Quoted by Allan Little, BBC News, June 2014.

145. F. de Waal (1998), *Chimpanzee Politics. Power and Sex among Apes*. Revised edition. Johns Hopkins University Press, Baltimore.

146. E. B. Keverne and J. P. Curley (2004), 'Vasopressin, oxytocin and social behavior'. *Current Opinion in Neurobiology*, vol. 14, pp. 777–83.

147. 'There are three conditions which often look alike / Yet differ completely, flourish in the same hedgerow: / Attachment to self and to things and to persons, detachment/ From self and from things and from persons; and growing between them, indifference/ Which resembles the others as death resembles life…' T. S. Eliot, 'Little Gidding'.

148. C. McCall and T. Singer (2012), 'The animal and human neuroendocrinology of social cognition, motivation and behavior'. *Nature Neuroscience*, vol. 15, pp. 681–8.

149. Oxytocin has been tried in cases of autism, characterized by lack of ability to form social or emotional bonds with others. So far, the results have not been very encouraging.

150. This poem was written by Wilfred Owen as a preface to a projected volume of his poetry. Subsequently, the last line was use on a commemorative tablet placed in Westminister Abbey to commemorate poets killed in the First World War. It is part of Poets' Corner, which commemorates large numbers of distinguished playwrights, authors, and poets (Chaucer was the first), in contrast to other parts of the Abbey, devoted largely to statesman and military leaders. The panel records the names of sixteen poets, among them Rupert Brooke and Wilfred Owen, who was killed seven days for the war ended.

151. 'The organizers of the anti-Vietnam War movement…would one day believe they had failed because the war, regardless of everything, continued for ten more years.…I wondered how many times a country could be disowned by a vital and intelligent sector of its youth before something broke, something deep inside its structure that could never be repaired again. The systole and diastole, the radicalization and the return of cautionary thinking, the bursts of idealism followed by equally quick swerves back to skepticism and the acceptance of things as they are—how many times before memory catches up with the latest swelling of the ideal and squashes it with cynicism before it can mature? In a word, how long is freedom? Is this the way America grows, or is this the way she slowly dies? Are these spasms of birth or of death?' Arthur Miller (1987), *Timebends. A Life*. Harper and Row, New York.

152. See *American Psychologist* (2013), vol. 68, part 7, for a series of articles about the psychology of war. Interestingly, none of them concerns itself with the special role of males in war.

153. For example, in the decisions of the US president and UK prime minister to invade Iraq in 2003.

154. B. J. Everitt and J. Herbert (1975), 'The effects of implanting testosterone propionate into the central nervous system on the sexual behaviour of adrenalectomized female rhesus moneys'. *Brain Research*, vol. 14, pp. 109–20.

155. See S. R. Davis (2013), 'Androgen therapy in women; beyond the libido'. *Climacteric*, vol. 16, pp. 18–24.

156. J. Bancroft and C. A. Graham (2011), 'The varied nature of women's sexuality: unresolved issues and a theoretical approach'. *Hormones and Behavior*, vol. 59, pp. 717–29.

157. P. Sapienza et al. (2009), 'Gender differences in financial risk aversion and career choices are affected by testosterone'. *Proceedings of the National Academy of Sciences USA*, vol. 106, pp. 15268–73.

158. P. A. Bos et al. (2012), 'Acute effects of steroid hormones and neuropeptides on human social-emotional behavior: A review of single administration studies'. *Frontiers in Neuroendocrinology*, vol. 33, pp. 17–35.

159. The field was really kick-started by E. E. Maccoby and C. N. Jacklin (1974) in their book *The Psychology of Sex Differences*. Stanford University Press.

160. D. Voyer, S. Voyer, and M. P. Bryden (1995), 'Magnitude of sex differences in spatial abilities: a metaanalysis and consideration of critical variables'. *Psychological Bulletin*, vol. 117, pp. 250–70.

161. Larry Summers, then President of Harvard University, resigned after a row about his stating that women in general achieved fewer high grades in mathematics because of sex differences in intrinsic ability. It remains true that there are far fewer women than men in University maths department, either as undergraduates or studying for a Ph.D.

162. W. Sommer et al. (2013), 'Sex differences in face cognition'. *Acta Psychologica*, vol. 142, pp. 62–73.

163. 'It is conjectured that human survival has depended to a large extent on accurate social judgements and that, as an evolutionary consequence, modular cognitive processes are devoted to such functions. Neuropsychological studies and human functional imaging provide partial support for this idea of a dedicated "social intelligence", particularly studies that address perception of facial expression.' J. S. Winston, B. A. Strange, J. O'Doherty, R. J. Dolan (2002), 'Automatic and intentional brain responses during evaluation of trustworthiness of faces'. *Nature Neuroscience*, vol. 5, pp. 277–83.

164. A. Herlitz et al. (2013), 'Cognitive sex differences are not magnified as a function of age, sex hormones, or puberty development during early adolescence'. *Developmental Neuropsychology*, vol. 38, pp. 167–79.

165. W. De Neys et al. (2013), 'Low second-to-fourth digit ratio predicts indiscriminate social suspicion, not improved trustworthiness detection'. *Biology Letters*, vol. 9, pp. 203–309.

166. see J. M. Heminway (2009), 'Female investors and securities fraud: is the reasonable investor a woman?' *William and Mary Journal of Women and the Law*, vol. 15, pp. 291–336. She uses the term 'reasonable investor' in much the same way as 'rational man' though with a few caveats. See Chapter 7 for more discussion on this topic.

167. This is explained more fully in Chapter 7.

168. Originally called the Stein–Leventhal syndrome, after its two discoverers.

169. There are some reports that PCOS is more common in female-to-male transgenders.

170. A. N. V. Ruigrok et al. (2014), 'A meta-analysis of sex differences in human brain structure'. *Neuroscience and Biobehavioral Reviews*, vol. 39, pp. 34–50.

171. As the previous chapter shows, the human female's brain also responds to testosterone.

172. His great textbook (*Cerebri anatome*) was illustrated by none other than Christopher Wren.

173. P. D. MacLean (1990), *The Triune Brain in Evolution. Role in Paleocerebral Functions*. Plenum Press, New York.

174. W. Cannon (1932), *The Wisdom of the Body*. W. W. Norton.

175. K. Woollett and E. A. Maguire (2011), 'Acquiring "the knowledge" of London's layout drives structural brain changes'. *Current Biology*, vol. 21, pp. 2109–14.

176. J. W. Papez (1937), 'A proposed mechanism of emotion'. *Archives of Neurology and Psychiatry*, vol. 38, pp. 725–43.

177. X. Xu et al. (2012), 'Modular genetic control of sexually dimorphic behaviors'. *Cell*, vol. 148, pp. 595–607.

178. Sir Godfrey Hounsfield was awarded a Nobel prize in 1979.

179. J. Ponseti et al. (2009), 'Assessment of sexual orientation using the hemodynamic brain response to visual sexual stimuli'. *Journal of Sexual Medicine*, vol. 6, pp. 1628–34.

180. R. J. R. Blair (2013), 'The neurobiology of psychopathic traits in youths'. *Nature Reviews Neuroscience*, vol. 14, pp. 786–99.

181. 'After the procedure [lesions of the amygdala]...the child is very quiet and obedient. Hair-cutting is now normally possible with no difficulty,

and even injections are accepted easily and with no force.... A visit to the department store or toy shop, taking him to the party in a friend's house, or even a trip by train are now performed with no special difficulty. The violent behavior and destructiveness in such child cases, though not so seriously antisocial as in the adult cases because of their age, also are much calmed.' H. Narabayashi (1972), 'Stereotaxic amygdalotomy'. In: *The Neurobiology of the Amygdala*, B. E. Eleftheriou (ed.), Plenum Press, New York, pp. 459–83.

182. S. C. Muller (2013), 'Magnetic resonance imaging in paediatric psychoneuroendocrinology: a new frontier for understanding the impact of hormones on emotion and cognition'. *Journal of Neuroendocrinology*, vol. 253, pp. 762–70.

183. A theme of many other disciplines, including Freudian psychology.

184. A recent neurological re-examination of this case is given by H. Damasio et al. (1994), 'The return of Phineas Gage: clues about the brain from the skull of a famous patient'. *Science*, vol. 264, pp. 1102–5.

185. A much fuller discussion of the role of the brain in criminality (and other factors, including genes and upbringing) can be found in A. Raine (2013), *The Anatomy of Violence: The Biological Roots of Crime*. Allen Lane.

186. A. L. Glenn and A. Raine (2014), 'Neurocriminology: implications for the punishment, prediction and prevention of criminal behaviour'. *Nature Reviews Neuroscience*, vol. 15, pp. 54–63.

INDEX

Bold entries refer to illustrations.

INDEX

vasopressin 90–1
vertebrates, and testosterone 19–21
viruses 31

war 131
  changed nature of 148
  Clausewitz's 'trinity' 132–3
  cooperative action 143
  differences between humans and
    chimpanzees 139–40
  distinction from intra-group
    aggression 141
  fanaticism 143–4
  frontal lobes' maturation 147–8
  group allegiance 133
  group fitness 134
  group identity 134, 135
  as male activity 134
  male bonding 134, 142, 143, 145
  moderation of tendencies towards 150
  neglect of biological and neurological
    factors 135
  non-human primates 135, 139–44
  primitive societies 137–8
  professional armies 139–40
  psychological factors 135
  rape 105–7
  social status of warriors 137–8
  street gangs 140–1, 142
  testosterone 143–4, 146–7, 148–9

uniforms 138, 139
young males 137, 138, 140, 141–2, 143, 144
  see also aggression
Wellington, Duke of 138
West, Rebecca 151
Willis, Thomas 174
Wilson, Edmund O. 3, 56, 72, 167
Winchelsea, Lady 151
women
  differences between men and
    women 159–61
  effects of testosterone 158–9, 162–3,
    164–5
  hormones and sexuality 152–7
  investment strategy 163
  menopause 63–4, 155
  menstruation 63, 153, **154**
  polycystic ovarian syndrome 164–5
  prenatal testosterone 163
  roles in finance 163
  testosterone and sexuality 64, 156–7,
    158, 163
  testosterone levels 152
Woolf, Virginia 151
Wynne-Edwards, V. 134

Yanomami 138
Young, William 40

Zuckerman, Solly 70–1

BAD MOVES

*How decision making goes wrong,
and the ethics of smart drugs*

Barbara Sahakian and Jamie Nicole LaBuzetta

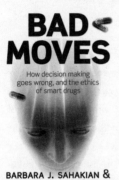

**BARBARA J. SAHAKIAN &
JAMIE NICOLE LABUZETTA**

978-0-19-966847-2 | Hardback | £14.99

"With this accessible primer, full of medical anecdotes and clear explanations, Sahakian and LaBuzetta prepare the public for an informed discussion about the role of drugs in our society." *Nature*

The realization that smart drugs can improve cognitive abilities in healthy people has led to growing general use, with drugs easily available via the Internet. Sahakian and LaBuzetta raise ethical questions about the availability of these drugs for cognitive enhancement, in the hope of informing public debate about an increasingly important issue.

BRAINWASHING

The science of thought control

Kathleen Taylor

978-0-19-920478-6 | Paperback | £9.99

"An ambitious and well-written study."

The Guardian

In *Brainwashing*, Kathleen Taylor brings the worlds of neuroscience and social psychology together for the first time. In elegant and accessible prose, and with abundant use of anecdotes and case-studies, she examines the ethical problems involved in carrying out the required experiments on humans, the limitations of animal models, and the frightening implications of such research. She also explores the history of thought control and shows how it still exists all around us, from marketing and television, to politics and education.

EXTREME

Why some people thrive at the limits

Emma Barrett and Paul Martin

978-0-19-966858-8 | Hardback | £16.99

"Deeply researched...the book is amusing, intriguing, exciting and a little horrifying."

New Scientist

"A rich and often compelling book."

Daily Mail

Why do some people risk their lives by placing themselves in extreme situations? What drives such people? What psychological and emotional qualities are needed by the successful deep-sea diver, mountaineer, astronaut, caver, or long-distance solo sailor? And are there lessons the rest of us can learn from them? In *Extreme*, psychologist Emma Barrett and behavioural scientist Paul Martin explore the challenges that people in extreme environments face, including fear, pain, loneliness, boredom and friction between individuals, and the approaches taken to overcome them. Using many fascinating examples, and drawing on the latest scientific discoveries, they show how we can all benefit from the insights gained.

HAPPINESS

The science behind your smile

Daniel Nettle

978-0-19-280559-1 | Paperback | £8.99

"A lucid and sensible survey of the latest research."   *Independent*

"Well written, accurate and engaging, with a lightness of touch that makes it a delight to read."
Nature

What exactly is happiness? Can we measure it? Why are some people happy and others not? And is there a drug that could eliminate all unhappiness? Daniel Nettle uses the results of the latest psychological studies to ask what makes people happy and unhappy, what happiness really is, and to examine our urge to achieve it. Along the way we look at brain systems, and mind-altering drugs, and how happiness is now marketed to us as a commodity. Nettle concludes that while it may be unrealistic to expect lasting happiness, our evolved tendency to seek happiness drives us to achieve much that is worthwhile in itself.

RUN, SWIM, THROW, CHEAT

The science behind drugs in sport

Chris Cooper

978-0-19-967878-5 | Paperback | £10.99

Drugs in sport are big news and the use of performance-enhancing drugs in sport is common. Cooper explains how these drugs work and the challenges of testing for them, putting into context whether the 'doping' methods of choice are worth the risk or the effort. Exploring the moral, political, and ethical issues involved in controlling drug use, Cooper addresses questions such as 'What is cheating?', 'Why do the classification systems change all the time?', and 'Should all chemicals be legal, and what effect would this have on sport?'.

Looking forward, he examines the recent work to study the physical limitations of rat and mice behaviour. He shows that, remarkably, simple genetic experiments producing 'supermice' suggest that there may be ways of improving human performance through genetic modification too, raising ethical and moral questions for the future of sport.

THE BRAIN SUPREMACY

Notes from the frontiers of neuroscience

Kathleen Taylor

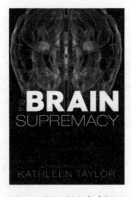

978-0-19-968385-7 | Paperback | £12.99

"The book shines in presenting a thorough and illuminating analysis of neuroscience methods, past and present. Taylor's explanation is thoughtful, engaging and provides readers with a valuable understanding of what different approaches can offer to both science and society as a whole."   *New Scientist*

Using recent examples from the scientific literature and the media, *The Brain Supremacy* explores the science behind the hype, revealing how techniques like fMRI actually work and what claims about using them for mindreading really mean. The implications of this amazingly powerful new research are clearly and entertainingly presented. Looking to the future, the book sets current neuroscience in its social and ethical context, as an increasingly important influence on how all of us live our lives.

Sign up to our quarterly e-newsletter **http://academic-preferences.oup.com/**

THE STRESSED SEX

Uncovering the truth about men,
women, and mental health

Daniel Freeman and Jason Freeman

DANIEL FREEMAN & JASON FREEMAN

the
stressed
sex

Uncovering
the truth about
men, women, and
mental health

978-0-19-965135-1 | Hardback | £16.99

Every day millions of people struggle with psychological and emotional problems. *The Stressed Sex* sets out to answer a simple, but crucial, question: are rates of psychological disorder different for men and women? The implications—for individuals and society alike—are far-reaching, and to date, this important issue has been largely ignored in all the debates raging about gender differences. In a finding that is sure to provoke lively debate, Daniel and Jason Freeman reveal that, in any given year, women experience higher rates of psychological disorder than men. Why might this be the case? *The Stressed Sex* explains current scientific thinking on the possible reasons—and considers what might be done to address the imbalance.